Recent titles in this series:

20. Hydrological maps. *Co-edition Unesco-WMO.*
21. World catalogue of very large floods/Répertoire mondial des très fortes crues.
22. Floodflow computation. Methods compiled from world experience.
23. Water quality surveys.
24. Effects of urbanization and industrialization on the hydrological regime and on water quality. Proceedings of the Amsterdam Symposium. October 1977/Effets de l'urbanisation et de l'industrialisation sur le régime hydrologique et sur la qualité de l'eau. Actes du Colloque d'Amsterdam. Octobre 1977. *Co-edition IAHS-Unesco - Coédition AISH-Unesco.*
25. World water balance and water resources of the earth. *(English edition).*
26. Impact of urbanization and industrialization on water resources planning and management.
27. Socio-economic aspects of urban hydrology.
28. Casebook of methods of computation of quantitative changes in the hydrological regime of river basins due to human activities.
29. Surface water and ground-water interaction.
30. Aquifer contamination and protection.
31. Methods of computation of the water balance of large lakes and reservoirs.
    Vol. I Methodology
    Vol. II Case studies
32. Application of results from representative and experimental basins.
33. Groundwater in hard rocks.
34. Groundwater Models.
    Vol. I Concepts, problems and methods of analysis with examples of their application.
35. Sedimentation Problems in River Basins.
36. Methods of computation of low stream flow.
37. Proceedings of the Leningrad Symposium on specific aspects of hydrological computations for water projects (Russian).
38. Methods of hydrological computations for water projects.
39. Hydrological aspects of drought.
40. Guidebook to studies of land subsidence due to groundwater withdrawal.
41. Guide to the hydrology of carbonate rocks.
42. Water and energy: demand and effects.

For details of the complete series please see the list printed at the end of this work.

Water and energy : demand and effects

Prepared for the International Hydrological Programme by the U.S. National Committee on Scientific Hydrology

George H. Davis

Unesco

Unesco wishes to gratefully acknowledge the contribution of the U.S. National Committee on Scientific Hydrology, under whose auspices the manuscript was prepared, and also the U.S. Geological Survey, Reston, Virginia, USA, in providing the camera-ready copy of this document.

Printed in 1985 by the United Nations
Educational, Scientific and Cultural Organization
7, place de Fontenoy, 75700 Paris

Printed by: Imprimerie Gedit, Tournai, Belgium

ISBN 92-3- 102328-4

Printed in Belgium

Preface

Although the total amount of water on earth is generally assumed to have remained virtually constant, the rapid growth of population, together with the extension of irrigated agriculture and industrial development, are stressing the quantity and quality aspects of the natural system. Because of the increasing problems, man has begun to realize that he can no longer follow a "use and discard" philosophy – either with water resources or any other natural resources. As a result, the need for a consistent policy of rational management of water resources has become evident.

Rational water management, however, should be founded upon a thorough understanding of water availability and movement. Thus, as a contribution to the solution of the world's water problems, Unesco, in 1965, began the first world-wide programme of studies of the hydrological cycle – the International Hydrological Decade (IHD). The research programme was complemented by a major effort in the field of hydrological education and training. The activities undertaken during the Decade proved to be of great interest and value to Member States. By the end of that period, a majority of Unesco's Member States had formed IHD National Committees to carry out relevant national activities and to participate in regional and international co-operation within the IHD programme. The knowledge of the world's water resources had substantially improved. Hydrology became widely recognized as an independent professional option and facilities for the training of hydrologists had been developed.

Conscious of the need to expand upon the efforts initiated during the International Hydrological Decade and, following the recommendations of Member States, Unesco, in 1975, launched a new long-term intergovernmental programme, the International Hydrological Programme (IHP), to follow the Decade.

Although the IHP is basically a scientific and educational programme, Unesco has been aware from the beginning of a need to direct its activities toward the practical solutions of the world's very real water resources problems. Accordingly, and in line with the recommendations of the 1977 United Nations Water Conference, the objectives of the International Hydrological Programme have been gradually expanded in order to cover not only hydrological processes considered in interrelationship with the environment and human activities, but also the scientific aspects of multi-purpose utilization and conservation of water resources to meet the needs of economic and social development. Thus, while maintaining IHP's scientific concept, the objectives have shifted perceptibly towards a multidisciplinary approach to the assessment, planning, and rational management of water resources.

As part of Unesco's contribution to the objectives of the IHP, two publication series are issued: "Studies and Reports in Hydrology" and "Technical Papers in Hydrology." In addition to these publications, and in order to expedite exchange of information in the areas in which it is most needed, works of a preliminary nature are issued in the form of Technical Documents.

The purpose of the continuing series "Studies and Reports in Hydrology" to which this volume belongs, is to present data collected and the main results of hydrological studies, as well as to provide information on hydrological research techniques. The proceedings of symposia are also sometimes included. It is hoped that these volumes will furnish material of both practical and theoretical interest to water resources scientists and also to those involved in water resources assessments and the planning for rational water resources management.

Contents

Page

Foreword

Rapid increases in world oil prices beginning in the early 1970s marked the start of a major transition in energy supply and use patterns. The rise in prices, together with generally declining oil and gas production, is expected to lead to extensive development of alternative energy sources that previously were too costly for widespread use. Chief among these alternatives are development of oil shale, tar sands, and coal for liquid fuels; coal as a source of gaseous fuel; and increased use of coal and geothermal energy directly for electric power generation.

Most of the alternatives require water--as a source of hydrogen in several processes and as a cooling agent in all. Many water-consuming projects have been proposed for areas that have no present surplus of water. Accordingly, there is widespread concern about whether available water supplies will be sufficient to meet present needs as well as additional water demands for energy development and whether other alternatives are available.

In response to concerns about the impact of energy development on water resources, the International Hydrological Programme (IHP) designated Mr. George H. Davis (United States) and Mr. A. L. Velikanov (USSR) as co-rapporteurs to prepare a report on "Hydrological Problems Arising from the Development of Energy Resources," with primary responsibility for drafting a preliminary report discussing the problems affecting water resources as a result of energy development, including water power generation, mining hydrology, geothermal energy, and pumped-storage systems, and emphasizing in particular the hydrological consequences of changes in methods of energy production. To this end the co-rapporteurs met at Unesco headquarters in Paris in July 1976 and prepared a preliminary draft report. During the ensuing year, substantive additions to the report were proposed by Mr. Velikanov, and Mr. Davis added some new material on Canadian tar sands development. After reviewing and editing the completed draft, Mr. Davis transmitted it to Unesco under covering letter of July 7, 1977.

In June 1977 the Intergovernmental Council of the IHP at its second session decided to expand the project on Hydrological Problems Arising from Energy Developments into a Working Group on "Study of the Hydrological Problems Arising from the Development of Energy Resources Including Water-Power Generation, Mining Hydrology, Geothermal Energy and Pumped-Storage Systems." The Working Group had the following terms of reference:

1. to study, on the basis of information received from individual countries and international organizations, the present and, more especially, the future use of water resources in the power industry, with the purpose of forecasting future water use;

2. to analyze methods of assessing the effects of the power industry on the quality of water resources;

3. to collect and analyze information on possible new methods of water use for energy production;

4. to prepare for publication a consolidated report on this subject;

5. to take into account in these areas the work carried out by international organizations such as the Organization for Economic Co-Operation and Development (OECD) and (ECE).

Four specialists were named to the group, designated Working Group 5.7 within IHP Project 5, "Investigation of the Hydrological and Ecological Effects of Man's Activities and Their

Assessment": Robert O. Ankrah, Ghana; George H. Davis, United States; Charles D. D. Howard, Canada; and A. L. Velikanov, the USSR.

Working Group 5.7 held its first session at Unesco House, Paris, from September 18 to 22, 1978, and elected as chairman George H. Davis. The group updated and completed the draft report prepared earlier by Mssrs. Davis and Velikanov, and published in 1979 as Unesco Technical Paper in Hydrology No. 17. The group then prepared an outline of the consolidated report requested by the IHP Council at its second session, including a timetable for its preparation and designation of responsibility for specific sections. Another task addressed by the Working Group was preparation of an inquiry to be addressed to National Committees for the IHP requesting information on present and future use of water resources in the energy sector with the purpose of forecasting future water use. Finally, the Working Group prepared recommendations for future activities in the second phase of the IHP.

The Secretariat of the IHP circulated the questionnaire drafted by the Working Group to all National Committees of the IHP together with the Final Report of the First Meeting of Working Group 5.7 requesting responses no later than July 1, 1979. Because few responses were received promptly, the Secretariat of the IHP circulated a second request in April 1979 and a further request in November 1979. The circulars requested general information from all member countries of the IHP and more detailed information on energy development from 28 countries. As of September 1980, technical input had been received from 6 countries, and only 3 (Canada, New Zealand, and Australia) had provided in-depth responses for the more detailed information requested.

In view of the inadequate response to the questionnaires, Mr. Davis proposed an alternate approach: using the available literature together with country inputs as a basis for the comprehensive report requested by the IHP Council. Due to funding limitations, Unesco was unable to support another meeting of Working Group 5.7 and no activities of the working group were budgeted by Unesco. In May, 1981, at a meeting in Reston, Virginia, of Mr. Davis with Dr. J. S. Gladwell, Division of Water Sciences of Unesco, and Dr. Della Laura, Executive Secretary of the U.S. National Committee on Scientific Hydrology, it was agreed that Mr. Davis should proceed with the new approach under the auspices of the U.S. National Committee on Scientific Hydrology, which is responsible for United States participation in the Unesco International Hydrological Programme.

1 Introduction

On a global scale, it is doubtful that energy development has a significant effect on water balances. The effects on water quality, however, appear to be more significant in the hydrologic cycle, as suggested by problems of intercontinental scope such as acid rain, changes in the CO_2 balance of the atmosphere, and release by nuclear power facilities of tritium and Kr^{85}, which enter the atmosphere and fall out in precipitation. (Atmospheric release and transport, involved in all three of these effects, are in the realm of meteorology rather than hydrology and, for that reason, are not treated in detail in this report.)

Despite the lack of global effects on water balances, the irregular areal distribution of energy and water resources and the pronounced time variability of the latter cause the effects of energy development to differ significantly throughout the world. Hence, most analyses take a regional approach.

Some water withdrawn in energy processes is incorporated in a product, such as synthetic gas, some may be dissipated to the atmosphere in evaporative cooling, while some is returned to surface or ground waters as a liquid. Accordingly, it is quite important to distinguish between water use (that is, gross withdrawal) and water consumption (that is, that part of the water not returned to the water system, termed "consumptive use"). Consumptive use is closely related to the general type of process and its overall thermal efficiency; accordingly, it may be estimated from production statistics. Gross water use, on the other hand, commonly is governed by cooling-system design, and therefore must generally be estimated on a site-by-site basis. Demand for water at a single site may also fluctuate. As Young and Thompson (1973) point out, with respect to electric power generation, the term "Water requirements" is misleading because demand for water for cooling is sensitive to the price of water delivered at the plant and thus is quite flexible rather than fixed as the term "requirements" implies. Much the same is true of other energy conversion processes, such as synthetic-fuel conversion, as pointed out by Probstein and Gold (1978).

Water plays a major role in many aspects of energy production and conversion, including mining and reclamation of mined lands, on-site processing and waste disposal, transportation, refining, and conversion to other, more convenient forms of energy (Davis and Velikanov, 1979). In areas where water supplies are generally adequate, most energy-related problems are in the realm of pollution and its adverse impacts on other water users. Where runoff is generally less than potential diversions, energy industries must compete with others users for the limited available supplies. Water supply is especially critical in areas receiving less than 250 mm annual precipitation, generally not enough for establishing vegetation without irrigation. Finally, in areas where the existing consumptive use approaches the perennial supply, any new water consumption imposed by energy development will be at the expense of existing uses.

In the following sections, discussions will be presented of the principal water-using processes in the energy sector, present water use and consumption by energy processes and estimates of future water use, the effect of energy development on water quality, new methods of water use in the energy sector, and alternatives available to energy planners faced with problems of limited water supplies or serious potential environmental impacts.

In this report much reliance is placed on developing world-wide estimates on the basis of extrapolation of experience in the United States and Canada by correlation with energy data compiled by the OECD, the United Nations, and the World Bank. The reasons for adopting this approach are threefold: (1) energy production statistics are available by individual countries world-wide and the statistics are detailed as to energy fuels and conversion process, and little comparable data are available on water use and consumption; (2) considerable information is available in the literature on water use and consumption by different energy processes in the United States and Canada; and (3) general consistency in design of large-scale energy processes throughout the world, such as electric power plants and oil refineries, permits extrapolation with a considerable degree of confidence.

The general pattern of discussion within the following sections is to treat extraction first and to follow the fuel through on-site processing, transportation, refining, and conversion to end use insofar as each stage is applicable and plays a significant role in water use. At this point a caveat is appropriate with respect to the matter of predicting future water use in the energy sector. This sort of prediction is extremely risky, more because of uncertainties in future energy growth rather than in the water aspects. To paraphrase an old quotation: "The road to hell is paved with abandoned energy scenarios."

Before launching into a discussion of water-using processes, a brief comment is appropriate on an energy source that requires water in its production but not in its general use--firewood. Wood has been used since the dawn of history for cooking and heating. As recently as the mid-18th century, wood supplied more than 90 per cent of the energy needs of the United States. By the 1970s, however, the contribution of wood to United States energy supplies had declined to less than 2 per cent of total energy consumption. With the rapid rise of energy prices in the 1970s, the use of wood as a fuel has been rediscovered, and the elegant term "wood biomass combustion" has been coined to describe it. Undoubtedly the burning of wood can be an important source of thermal energy even in the highly developed countries, especially in forest areas, where the waste from forest-product industries can be put to economic use to produce industrial and space heat.

In the developing countries, wood plays a more significant role in national energy budgets. In the low-income countries (those with per capita Gross National Product below US $360 per year), wood and agricultural or animal wastes furnish more than half the total energy supplies (World Bank, 1980). These locally produced, non-commercial fuels account for virtually all the fuel used in rural areas and about 25 per cent of total energy consumption in the developing countries. Some 2 billion people currently use these traditional fuels for cooking (World Bank, 1980, p. 55). Most of these people have access to charcoal, but between 500 million and 1 billion people must rely on agricultural and animal wastes for their cooking fires. It is one of the great tragedies of our time that the real price of alternative fuels has been rising, while the poorest nations have been consuming their wood supplies far more rapidly than they can be renewed. In an overall energy strategy, measures are needed to address the energy requirements of rural households as part of a development strategy that includes reforestation and the planting of trees on marginal land in cultivated areas.

Metric units are used generally throughout the report, and, in keeping with the practice of the OECD, large quantities of energy fuels are expressed in millions of metric tons oil equivalent (Mtoe). Electrical generation and other electrical quantities are given in kilowatts (kW) and gigawatts (GW), where $1 \text{ GW} = 1 \times 10^6 \text{ kW}$. A conversion table is included as Appendix B giving metric units and their English equivalents and other conversion factors for quantities used in this report. Specialized terms are defined in the glossary, Appendix A.

2 Principal water-using processes in the energy sector

The principal water-using processes in the energy sector, in order of total withdrawals, are (1) hydroelectric power generation, (2) steam-electric power generation, and (3) oil refining. Most other withdrawals in the energy sector are minor in comparison. If we take into account the distinction between water use and water consumption, hydroelectric power generation is generally considered to be of little consequence as a consumptive use because water is passed through hydroelectric turbines with no significant change in physical or chemical character. Although evaporation from man-made reservoirs is a major element of water consumption in many countries--because reservoirs serve many purposes other than hydroelectric generation, including flood control; irrigation, municipal, and industrial water supply; and recreation--it is difficult to estimate on a rational basis how much evaporation is properly chargeable to hydroelectric power generation.

As discussed in Chapter 3.0, the largest consumers of water in the energy sector in order of total world consumption are (1) steam-electric power generation, (2) oil refining, (3) oil and gas production (chiefly for water flooding), and natural gas processing. The impacts of these consumptive uses is tempered somewhat because much of the water consumed is saline: sea water in the case of oil refineries and steam-electric plants and formation water in the case of oil production.

Steam-electric power plants generally are sited close to electric load centers so as to minimize transmission-line losses. However, many of the large load centers are close to the sea and a large part of the steam-electric generation is cooled by sea water or by estuarine waters. Although they are not classified as saline, estuarine waters are not suitable for many water uses, and thus their use does not deplete consumable fresh-water resources.

Oil refining, on the other hand, is generally done at convenient sites along transportation routes from producing to marketing areas. Because most of the world's oil trade travels in marine tankers, the most convenient locations for refineries commonly are seaports; among these are such centers as Rotterdam, Houston, Los Angeles, New York, Aruba, Trinidad, Singapore, and Abadan.

In oil and gas production the major consumptive uses are for water flooding (secondary recovery), which is widely practiced throughout the mature oil fields of the world, and for enhanced oil recovery (tertiary recovery), which includes steam flooding and miscible-fluid flooding, both of which require large volumes of water. The use of water in oil production is somewhat different than most other consumptive uses in that water is used to flush oil out of the producing formations rather than being dissipated to the atmosphere. However, because the water so used is permanently removed from the active hydrologic cycle, it is classified here as a consumptive use. Where saline water is availiabe (such as formation waters produced with oil), it is used in water flooding in preference to fresh water, but in most water-flooding projects this makes up a relatively small part of the water requirement.

The major consumptive uses in natural-gas processing are for compressor-station cooling and for gas-cleaning facilities. For the most part, fresh water is used in these instances. Finally, the water needed to operate coal-slurry pipelines (about equal volumes of coal and water) can be thought of as a consumptive use, although little water is removed from the fluid phase of the hydrologic cycle. Nonetheless, such water is lost to the area of origin of the pipeline and represents to other water users an unreplenished withdrawal comparable to consumptive use.

2.1. EXTRACTION

2.1.1. Coal mining

Coal assumed importance as an energy source with the beginning of the industrial revolution in Europe. In the United States, the coal industry began with mining of bituminous coal in Virginia and anthracite in Pennsylvania. Production increased steadily in the industrialized countries during the 19th century, and by 1900 coal was the dominant source of energy for space heating, as a source of coal gas, for steam generation, and for coke in steel production. At that time coal supplied 90 per cent of energy consumption in the United States and comparable levels in Europe.

During the period 1900-1950 coal consumption grew less rapidly than total energy production because more convenient and competitively priced natural gas became available and new uses of oil, especially in the transportation sector as liquid fuel, increased rapidly. By 1950 coal made up only 38 per cent of the United States energy supply, and in 1972 it reached a low of about 20 per cent.

Coal's declining role in the United States energy structure was accelerated by government encouragement of development of nuclear energy, encouragement of imports of inexpensive foreign oil, and the maintenence of low prices for natural gas through price regulation. Also, the drive for improvement of air quality in the late 1960s and early 1970s tended to cause industrial coal users to seek cleaner burning fuels. These trends were mirrored in most of the industrial countries of the world to varying degrees, although in Europe the leading coal producing nations, the Federal Republic of Germany and the United Kingdom, made strenuous efforts to maintain domestic coal production for social and economic reasons. Unlike the United States, which for much of this century had enjoyed a surplus of natural gas, the local natural-gas supplies were small and the international traffic in natural gas had not yet begun. However, with the development of the huge Groningen Gas Field in the Netherlands in the 1950s and the North Sea oil and gas fields not long after, the United Kingdom and the Western European nations rapidly expanded their use of natural gas for space heating and industry. The Western European nations now import significant quantities of natural gas via pipeline from the USSR and from North Africa via sea-going tankers.

Coal suffered great market erosion to these alternative fuels at the same time that increasing concern (and subsequent Federal Regulations) for the health and safety of miners and for protection of the environment were reducing productivity and increasing the cost of coal. The only market segments in the United States where coal was not displaced were markets for metallurgical uses and for electric-power generation, where rapid, continuing growth made possible stabilization of coal demand despite coal's loss of market share. Coal production in the United States reached a low of less than 400 million metric tons per year in 1960 but then increased gradually to a level of about 540 million metric tons in 1972. In part at least, this was due to the far-sighted planning of electric utilities, which, foreseeing the coming rise in petroleum prices, maintained a considerable base of coal-fired generating capacity.

With rapid increases in world oil prices beginning in 1973, the demand for coal as an alternative fuel increased sharply, and this was reflected in increases in coal prices and production. In the United States, production increased at a rate of 6 to 7 per cent annually and reached 742 million metric tons in 1981. Comparable production gains have not been realized in Europe, but Australia and South Africa, reflecting increased world demand, have recorded rapid increases in coal production, producing 81 and 76 million metric tons, respectively, in 1979 (OECD, 1982b).

Coal production is divided into two basic categories: underground and surface mining (Figure 2.1.1). The selection of a mining technique depends mainly on the thickness of coal and the thickness and character of the overlying materials. Although surface mining is the predominant production method in the United States, accounting currently for 55 per cent of the total production, 68 per cent of the reserves are accessible only by underground mining.

2.1.1.1. Underground mining. The major underground mining techniques currently used in the United States are room-and-pillar and panel mining. In Europe, where mining is generally at greater depth than in the United States, panel mining is generally preferred. Room-and-pillar mining currently accounts for more than 90 per cent of United States production, but panel mining is expected to make up an increasing proportion of the total in the future.

Room-and-pillar mining involves removal of coal from intersecting tunnels penetrating the coal seam; coal pillars are left in a rectangular pattern to provide support for the roof and overlying deposits. Recovery in the room-and-pillar system averages about 55 per cent, but

Figure 2.1.1. Black Mesa coal mine in northeast Arizona, USA. Dragline, right center, is stripping overburden to expose coal and casting overburden on conical heaps to left. Farther left are regraded and reclaimed land. Coal is trucked out on haul roads, as at left center. This mine is dedicated exclusively to supplying Navajo and Mojave power plants (Figures 2.4 and 7). (Photo by U.S. Bureau of Reclamation.)

as the depth of mining increases the proportion of coal that must be left to support the roof increases and ultimately reaches a point where complete recovery techniques become more practicable. The principal steps involved in room-and-pillar mining are undercutting the seam, drilling blastholes, blasting, and loading and removing the coal. Roof bolts are used to maintain the structural integrity of the roof, and ventilation is advanced as mining progresses (Figure 2.1.1.1, block A). In the past these tasks were done by hand, but in recent decades they have been largely mechanized. Currently, most room-and-pillar mining is done with continuous-mining machines which combine the four traditional steps into a single operation. The continuous miner uses a rotating-drum head with replaceable teeth to excavate the coal, which is then loaded into shuttle cars (Figure 2.1.1.1, block B).

Panel mining differs from room-and-pillar mining in that the need to retain coal pillars for roof support is eliminated. Although limited to favorable geologic conditions and areas where immediate subsidence of the overlying surface can be tolerated, resource recovery can be increased to 85 per cent through panel mining. The preferred method of panel mining is the longwall technique, in which continuous mining machines are used to cut two parallel tunnels into the coal seam up to 1.5 km long and 200 m apart. These tunnels are used for ventilation, access, and coal removal. The panel of coal between the two tunnels is removed by a shearing machine that employs a rotary head to cut a slice of coal up to 1 m thick from the exposed face. The coal is removed from the working face by a continuous conveyer. Self-advancing hydraulic roof supports, which protect the machinery and operators, move forward to a new position as the shearing machine traverses the face, and the unsupported roof over the mined-out area is permitted to collapse (Figure 2.1.1.1, block C).

This method results in immediate rupture of the overlying strata and leakage downward of ground water from any aquifers in the overburden. Where significant aquifers overlie the coal, panel mining can have a severe adverse impact on ground water resources. However, the alternative of room-and-pillar mining is not necessarily more environmentally desirable over the long term, as pillars tend to collapse over time and the resulting collapse and land subsidence phenomena are prolonged, for many decades in some cases. Moreover, where mines are above the gradient of nearby streams, the abandoned mine tunnels, generally partly collpased, commonly act as subsurface drains, discharging large volumes of degraded water to nearby streams. Indeed, it is estimated that of the 16,000 km of streams in the eastern United States severely degraded through acid drainage and related chemical problems (Appalachian Regional Commission, 1969), three-quarters of the problem can be attributed to underground mining (Johnson and Miller, 1979).

Longwall mining is the preferred method in Europe, where seams currently mined are generally much deeper than in the United States and the weight of the overburden is proportionally greater. In the United Kingdom, for example, 95 per cent of current mining is by the longwall method. Shortwall mining, a variation of panel mining, named for the shorter panel face worked, generally on the order of 50 m, uses continuous mining machines in combination with hydraulic roof supports. This method has gained favor in the United States because it requires less capital and is more flexible than longwall mining. However, productivity is sacrificed because shuttle cars in lieu of continuous conveyers must be used to transport the coal.

Research is currently under way on hydraulic mining systems in which remotely controlled, high-pressure water jets are used to break the coal loose; the resulting coal-water slurry is then pumped to the surface, where the coal and water are separated. Should this technology gain wide acceptance, it could have important water-resources implications because large amounts of water would be needed for the circulating systems. However, makeup water demand is poorly known; with extensive recirculation of water it would be possible to keep makeup demand at a reasonable level.

2.1.1.2. Surface mining. Surface production of coal is carried out by the area, contour, auger, and pit methods, depending on the thickness and attitude of coal seams and overburden and the topographic relief of the terrain. Area mining, which is extensively used in flat terrain, as in the midwestern and western parts of the United States, entails removal of overburden by cutting successive trenches, typically 30 m wide and 1.5 km long, to expose the coal seam (Figure 2.1.1.2). The coal is then excavated and loaded into trucks through the use of large power shovels. Following coal extraction, another parallel trench is excavated and the overburden material is placed in the previously excavated trench. The irregular rows of spoil are graded to approximate the original terrain, and topsoil, which was segregated in the original excavation work, is spread and seeded to provide a cover of vegetation which minimizes erosion. Recovery in the area mining system typically is 80 to 90 per cent of the coal reserves. For many years the ratio of overburden to coal thickness that was economic to mine was about 10 to 1; however, with increased coal prices in recent years and the development of more efficient earth-moving equipment, ratios up to 60 to 1 are now economically feasible.

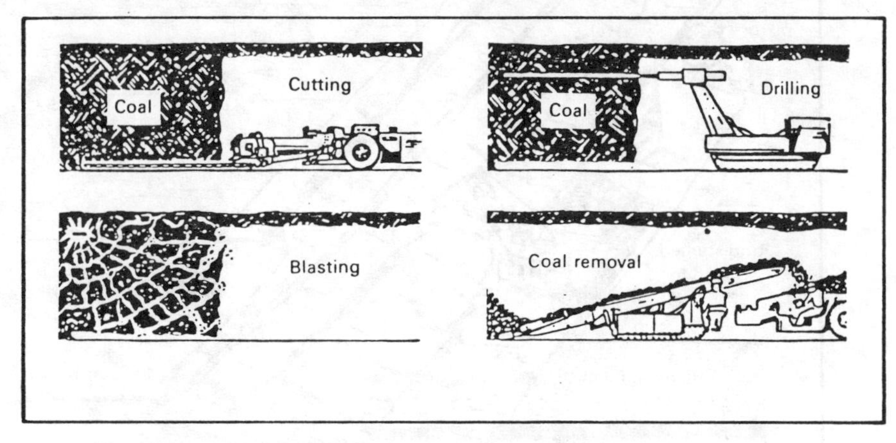

A. – Conventional mining.

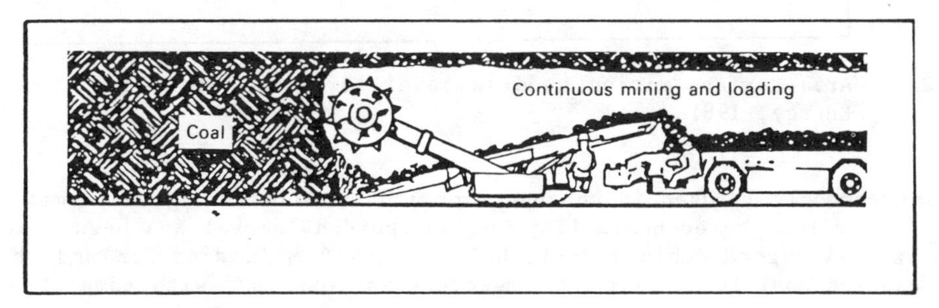

B. – Continuous mining.

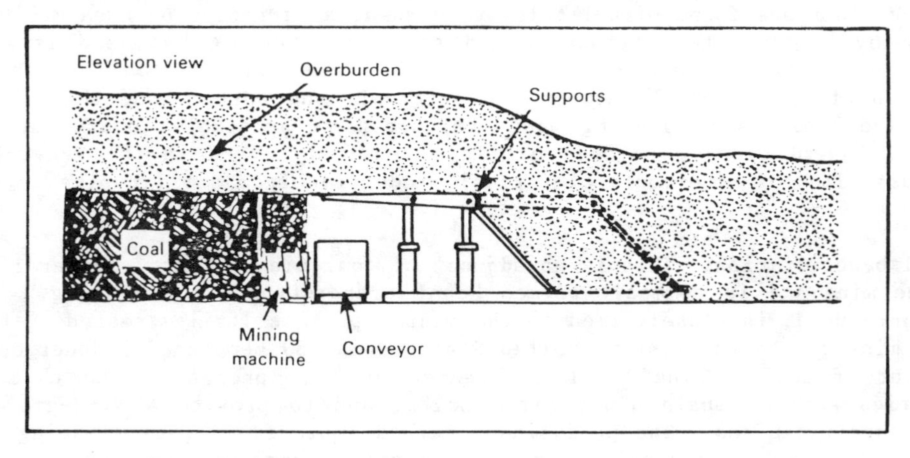

C. – Longwall mining.

Figure 2.1.1.1. Principal types of underground coal mining: (a) Conventional techniques used in room-and-pillar mining, (b) Continuous miner, (c) Longwall mining. (Source: U.S. Department of Energy, 1981a.)

Contour mining is practiced where near-horizontal coal beds crop out along the sides of mountains, as in the Appalachian region of the United States. The overburden is removed with excavating machines, beginning at the outcrop. As the exposed coal is removed, additional cuts are made in the slope until increasing overburden thickness makes further mining uneconomic. When such a limit is not reached before excavating to the top of the slope, the entire top of a mountain may be removed and the full coal seam mined out. In the past, spoil commonly was disposed of by pushing it to the outer edge of the excavation and letting it slide down the downslope. However, current law in the United States prohibits that practice, so, to minimize handling costs, great ingenuity must be employed in moving and storing the overburden until the reclamation stage.

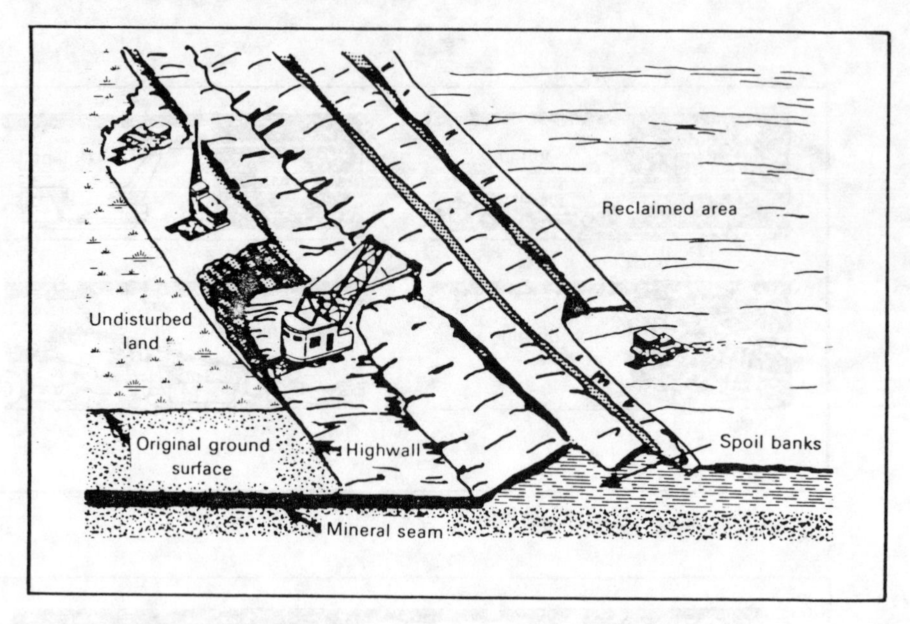

Figure 2.1.1.2. Area surface mining used in level terrane. (Source: U.S. Department of Energy, 1981a.)

Auger mining commonly is used in conjunction with contour mining to permit further mining from the coal face after the economic limit of overburden removal has been reached. In this technique, horizontal auger machines drill holes up to 2 m in diameter and as much as 60 m deep into the exposed coal face. Although recovery is less than with complete removal of the seam, as much as 30 to 50 per cent of the coal within reach of the augers can be recovered before the face is covered and the site is recontoured and revegetated.

Where thick seams are found with little overburden, a variation of area mining by open-pit methods is employed. Here the overburden and soil are stripped back and large shovels are used to excavate the face in a broad, expanding pit. Because the coal commonly makes up a large proportion of the material between the original land surface and the base of the coal seam, reclamation consists of grading what overburden is available toward the pit so as to provide minimal inward slopes. Brown coal is extensively mined in the Federal Republic of Germany and Australia for mine-mouth electrical production from similar pits as much as 200 m deep.

2.1.1.3. Coal beneficiation. This is an adjunct of coal mining that is generally carried out at or near the mine and is commonly termed "coal preparation" or "cleaning." Because this technology, where used, is closely tied to the mining process, it is treated in this report as an element of mining. In the eastern United States about 60 per cent of undergound-mined coal and 25 per cent of surface-mined coal is treated by this process. The objectives are to reduce impurities such as shale and pyritic sulfur and to provide a size-graded product to meet customer specifications. The predominant method employed in coal cleaning is the use of jigs, shaker tables, and high-density fluids to provide gravity separation of the relatively light coal from heavier mineral and rock which has about twice the density of coal. The fluids commonly employed are suspensions of clay in water, in which the density can be adjusted by the addition of iron oxide, barite, or other dense mineral matter. The major reasons for coal beneficiation are to improve its heating value and to remove impurities such as sulfur that can result in air pollution. However, coal-cleaning plants have themselves resulted in water pollution, in some cases, through release of water containing finely divided coal particles and through leaching of the plant wastes. As shown in Appendix C, the water consumption attributed to coal-cleaning processes is estimated to exceed that of mining a comparable weight of coal.

2.1.1.4. Water use. The principal uses of water in coal mining are for dust control in both surface and underground mining, for revegetation of surface-mined lands in arid and semi-arid regions, and in coal-cleaning. Water is used liberally to minimize airborne dust in undergound mines because coal dust is a serious health hazard to the miners. Similarly, water is used extensively in surface mining to minimize the airborne transport of dust on and off the mine site. Sprayers are used during active excavation work, and haul roads are wetted frequently to keep dust down.

Estimates of total water use in coal mining are crude at best because the amounts of water used are relatively small and commonly are supplied from mine drainage. These estimates cannot be relied upon for planning but are of use in judging the relative magnitude of water use in mining. Because recirculation is used widely in underground mining and little water is returned to the surface, the gross water use can be assumed to represent consumption. Similarly, in surface mining, essentially all the water used for dust control evaporates and thus is appropriately classified as consumptive use. Water used in coal beneficiation is not readily accounted for, although some evaporation would be expected from the continuously circulating clay slurries. There is little or no heat dissipated in the process, so the amounts of water ascribed to consumptive use are most likely a combination of modest evaporation and minor leakage to surface and ground water from the circulation systems.

Consumptive use of water for revegetation of reclaimed lands in arid and semi-arid areas is of little concern except in water-deficient areas, such as the western United States, where vast coal reserves occur in areas of less than 250 mm annual precipitation. The unit water demand for this purpose is highly variable, depending on ambient precipitation and the area disturbed per ton of coal mined. Locally, as much as 300,000 metric tons of coal are mined per hectare disturbed. In such places the amount of water required for revegetation becomes a trivial figure in terms of liters per ton mined.

2.1.2. Oil and gas extraction

Although petroleum from surface seeps had been used as a lubricant and sealant since the dawn of history, widespread use of oil as a fuel dates to the mid-19th century, following the drilling of the first commercial oil wells in Pennsylvania, USA, dating from 1859. For the 30 years that followed, the United States was virtually the world's only source of crude oil and refined products, the principal product being illuminating oil. By the turn of the century, United States oil production had reached nearly 8 million metric tons annually. With the development of the internal-combustion engine and the growth of the use of automobiles and trucks, oil found a ready market in the transporation sector, and it soon displaced coal for raising steam for boilers in many uses, because it was both easier to store and handle and less costly on an equivalent energy basis. Oil production grew in many countries to meet the growing demand as the world economy shifted from coal to the more convenient new fuel. By 1930, major amounts of oil were also being produced in the USSR, Mexico, Indonesia, Romania, Venezuela, and Iran; the oil was moved for the most part by marine tanker to refineries located in the major consuming areas.

Oil occurs in liquid form in porous marine sediments. It originates as a breakdown product of marine organic matter buried with the sediments. Being lighter than water, which occupies most of the pore space of sedimentary rocks, the oil tends to migrate upward and to accumulate in locations where it is trapped by impermeable rocks at high points of geologic structures. The prinicpal task of the oil explorer is to pinpoint such traps for testing by drilling. Oil, once located, is extracted through wells (Figure 2.1.2). The permeability of the oil-containing rocks controls the rate of flow toward wells and the proportion of the oil in-place that can be produced. Some formations, although known to contain oil, yield it at such a slow rate that their development is uneconomic. Generally, the initial production of an oil well represents the maximum rate at which the well will produce. As the pressure driving the oil toward the well is depleted, the production of the well declines at a rate determined largely by the permeability. Thus, to maintain a given level of production for a nation, it is essential to continue exploring for new deposits and to maintain a vigorous program of bringing new wells into production.

For over a century the United States was the world's leading producer of oil, supplying its own demand and exporting the surplus production. However, large new discoveries, mainly in the Persian Gulf area, following World War II resulted in a huge world surplus of producing capacity which depressed oil prices for two decades. This encouraged great increases in the use of oil, thus bringing world supply and demand into rough equilibrium by about 1970. From 1950 through 1970 world production increased exponentially from 2 billion to 17 billion barrels annually. The discovery rate of new fields had been declining sharply since the late 1950s despite active exploration efforts, and the United States found itself in the position of having to increase its oil imports to meet the ever-growing demand. With control of the market assured, a number of the leading oil-producing nations formed OPEC (Organization of Petroleum Exporting Countries) with the clear purpose of controlling oil production to force a rise in world oil prices. Beginning with a sharp increase in price in 1973, the cost of crude oil rose from about US $3 per barrel in the early 1970s to more than US $30 per barrel in

Figure 2.1.2. Typical offshore drilling platform, Gulf of Mexico, USA. Three-legged supports rest on sea bottom during drilling. Structure is moved by jacking-up legs and floating it on barge-like understructure. (Photo by U.S. Bureau of Mines.)

1981. The effects of this rapid price escalation have had a major impact on world economics and economic growth. In every country there has been a frantic scramble to increase domestic oil production, to conserve the now costly oil wherever possible, and to shift essential energy consumption to alternative, cheaper fuels. An even more widespread effect has been a pronounced slowing of economic growth rates throughout the world, and as of the close of 1982, the entire world was in the grip of an economic recession. With dropping demand, the existing producing capacity far exceeds the demand.

Natural gas is similar to oil in that its origin and occurrence are closely related, gas representing the lighter hydrocarbon products of the same processes that result in oil formation. Indeed, a large proportion of natural gas production is as a by-product of oil production; the term "associated gas" is used to distinguish that gas from gas which occurs without oil being present. Petroleum spans a spectrum ranging from solid bitumen to heavy (highly viscous) oils to light oils to natural gas liquids (LNG), associated gas, and finally to so-called dry gas with minimal fluid hydrocarbon content. In general, this gradation is defined by the ratio of carbon to hydrogen in the organic molecules, a higher proportion of hydrogen resulting in more mobile components. Because the distinctions are gradational, the terminology used in the petroleum trade is necessarily arbitrary, but it is used uniformly on a world-wide basis.

A major distinction between oil and natural gas that greatly affects the value and utility of gas is that, being in the gas state, natural gas is more difficult to store and transport than oil.

The first natural gas well in the United States was put into production at Fredonia, New York, in 1821. The discovery of oil nearly four decades later led to the discovery of large reserves of associated natural gas for which there was no market at the time. Thus, gas produced with oil was wasted by flaring. However, once the possible uses of natural gas for space heating, cooking, lighting, and raising steam were recognized, it quickly displaced gas manufactured from coal in many roles. The first large-scale use of natural gas for industrial heat was in the manufacture of glass and steel in western Pennsylvania, USA. Initially, the use of natural gas was confined to areas near producing oil fields, but with the development of long-distance gas transmission lines in the 1930s, gas quickly displaced coal a source of space heat and industrial heat. The amount of gas marketed in the United States, grew from a little more than 1×10^{12} ft^3 (23 Mtoe) in 1935 to 8×10^{12} ft^3 (186 Mtoe) in 1952, and continued to grow at an annual rate of 6.5 per cent over the next two decades, reaching 21×10^{12} ft^3 (490 Mtoe) in 1970 and peaking at nearly 22×10^{12} ft^3 (510 Mtoe) in 1973. As oil prices escalated after 1973, gas prices rose at a somewhat slower rate, resulting in conservation efforts which stabilized gas consumption in the United States at about 20×10^{12} ft^3 (465 Mtoe) per year through 1980.

The principal natural gas producing nations other than the United States are the USSR (333 Mtoe in 1979), the Netherlands (72 Mtoe in 1979), Canada (70 Mtoe in 1979), the United Kingdom (34 Mtoe in 1979), and Romania (33 Mtoe in 1979). This production in no way represents the world's potential gas production, but rather reflects the difficulty of transporting it in international trade. As of 1979, the principal international gas traffic was by pipeline from the Netherlands, Norway, and the USSR to central and western Europe, and from Canada to the United States. Marine transport was limited to exports from Algeria to western Europe and the United States totaling 10 Mtoe per year, and imports by Japan from Indonesia, Malaysia, the United Arab Emirates, and the United States totaling 19 Mtoe per year. With world oil production apparently leveling off at the current rate for the next 20 years, it is expected that natural-gas production and consumption will grow at a steady rate, with natural gas continuing to make up about 20 per cent of the world energy supply through the turn of the century.

Production of natural gas has much in common with production of oil. Natural gas is produced by tapping deposits trapped in porous sediments in geologic structural highs, and the techniques of exploration and production are quite similar, with allowance for the difference in physical state. As in the case of oil, the rate of production of a gas well declines from an initial high level, and, to maintain a given level of production, it is necessary to bring new wells into production on a regular schedule.

With respect to the use of water and the impact on water resources, however, oil and gas are distinctly different. Oil production results in the production of varying amounts of saline formation waters, which must be disposed of properly to avoid serious pollution of water resources. Gas production, on the other hand, results in essentially no accessory water production. Gas production, beyond an initial need for water for drilling fluid, requires no water (although processing and compressing gas for transportation involves substantial water use, as will be discussed later). Oil production, on the other hand, entails substantial water use in the secondary and tertiary stages of recovery.

Oil production is divided into three distinct phases. When a field is first developed, the oil commonly flows at land surface, the pressure being provided by gas in solution in the oil. This pressure drive is subject to normal decline as the reservoir is depleted, and the wells cease to flow. At that point, pumps are installed and the oil is lifted to land surface with mechanical energy, gravity providing the energy to move the oil toward the wells. This, too, is subject to depletion. Finally, the gravity flow drops to a level at which continued production is uneconomic. At this stage in the depletion of a reservoir, various options are available for continuing to produce. Gas can be injected through injection wells to increase the driving force; this commonly is done early in the life of the project to maintain flow as long as possible. The most widespread practice, however, is water flooding, in which water is injected into the producing formation at carefully spaced injection wells to displace the oil physically and drive it laterally toward producing wells. Depending on the permeability distribution of the producing zone, a given formation may or may not be amenable to this methodology. Water flooding, or "secondary recovery" as it is termed, is widely practiced in the mature oil fields of the world. Indeed, one-third of current United States and Canadian oil production is estimated to be from water-flooding operations. Primary production and water flooding generally result in recovery of about one-third of the oil originally present. With tertiary recovery, up to half the oil originally present may be recovered. Nevertheless, almost half the oil in-place is still unrecoverable with current technology.

2.1.2.1. Primary recovery. This term comprises all methods of recovery that rely on natural-drive mechanisms or the use of pump lift. The principal method of increasing primary production is drilling additional wells at closer spacing than the original grid.

2.1.2.2. Secondary recovery. This term encompasses methods of injecting water, gas, or air into the producing zone to maintain reservoir pressure and thus drive oil toward recovery wells. In modern practice, secondary recovery techniques may be used quite early in the life of a field to prolong flow as long as possible and thus avoid the additional cost of pumping. Water injection is the principal method of pressure maintenance under present economic conditions. With this technique, specially designed injection wells are drilled in a pattern to optimize subsequent recovery from normal production wells. Formation water produced with oil or from chemically compatible zones is preferred for the injection supply so as to minimize clay-water reactions that can impair the permeability of the producing zone. Where supplies of saline water are inadequate, treated freshwater is used. Injection rates and pressures are highly variable and depend on the character of the producing zone and the economics of the project. A principal environmental consideration is the avoidance of excessive injection pressures, which can result in damage to the casing and the producing zone and, in some situations, can trigger earthquakes.

2.1.2.3. Tertiary recovery. This encompasses all other techniques to increase the recovery of in-place oil not recovered using primary and secondary methods. The major methodologies include steam injection, in-situ combustion, miscible-gas injection, and micellar-polymer injection. Steam injection, the oldest of the methods, has been used commercially since 1961. It accounted for about three-fourths of the 15×10^6 metric tons per year of United States production by tertiary processes in 1980 and about 3.6 per cent of total oil production. The only other countries reporting significant production by this technology were Venezuela, with 7×10^6 metric tons per year, and Canada, with 9,000 metric tons per year (Crouse, 1981).

The main thermal processes used are steam soak and steam drive. The steam-soak technique increases production by heating the oil to reduce its viscosity so that it will flow more readily toward the well. This method has proven especially useful in the steeply dipping producing zones of California. Cyclic steam soak requires no special injection wells. Steam is injected into the reservoir for a period of 1 to 2 weeks. The wells are then closed off to allow the steam to penetrate and heat the formation. Oil and water are then pumped for several months until production declines below an acceptable level. The cycle may then be repeated. An estimated yield of 20 to 35 per cent of the oil in-place has been achieved using this method. The process is applicable to both large and small reservoirs and has the advantage that continuous generation of steam is not required.

In the steam-drive process, steam is injected continuously into the producing zone to advance a steam zone laterally through the reservoir and displace oil and water to producing wells. The process involves complex interactions of chemical and physical phenomena including, mainly, viscosity reduction and steam distillation. In a simplistic way, it can be envisioned as the advance of heat and oil-displacement fronts. The steam-drive method has several limitations. Because of rapid heat loss from surface equipment and the well bore, the method is limited to fairly shallow reservoirs, within about 2,000 m of land surface, and sweep efficiency is often less than satisfactory (Figure 2.1.2.3A).

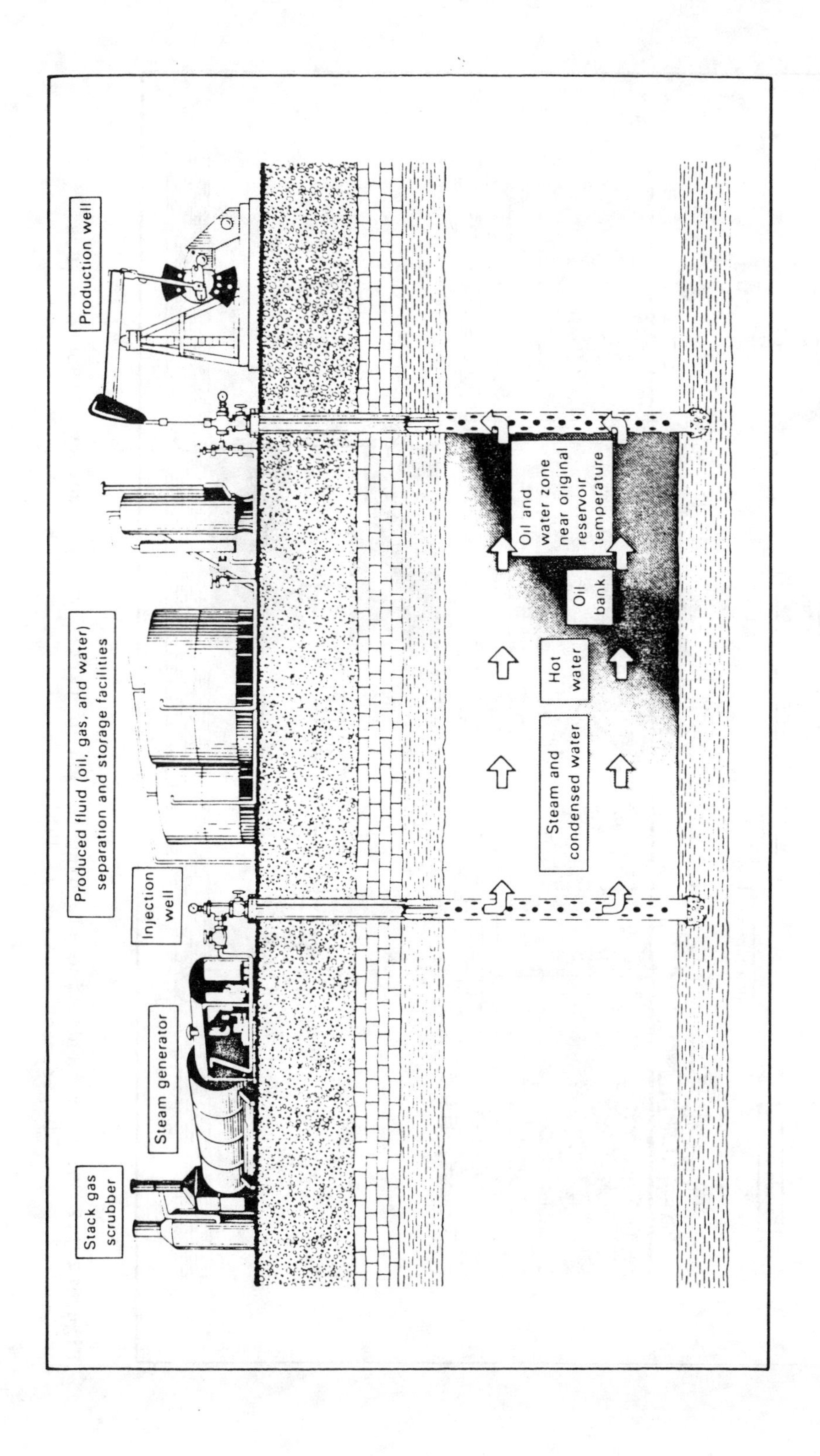

Figure 2.1.2.3A. Steam-drive method of tertiary oil recovery. (Source: U.S. Department of Energy, 1981a.)

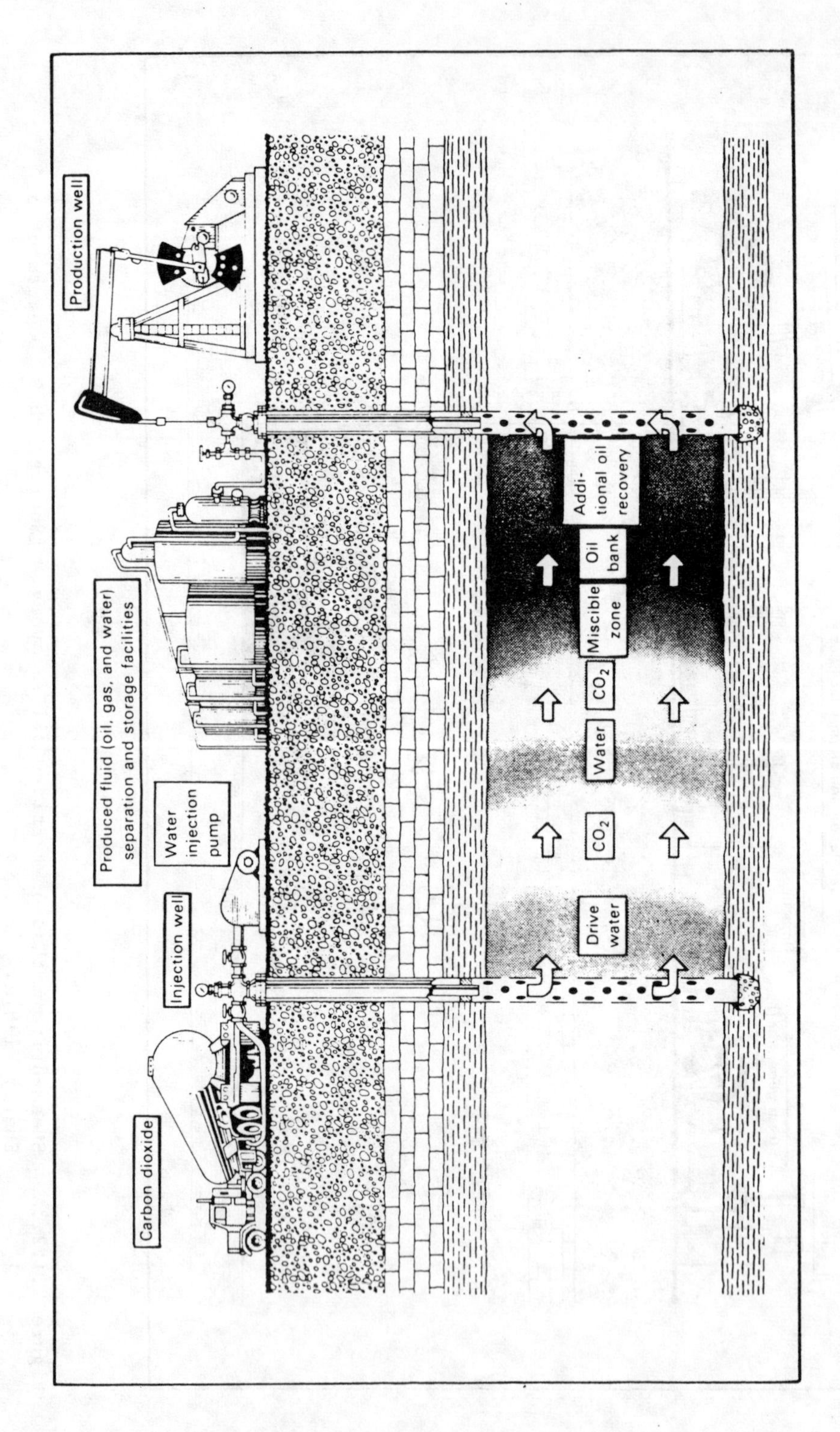

Figure 2.1.2.3B. Carbon dioxide flooding method of tertiary oil recovery. (Source: U.S. Department of Energy, 1981a.)

The in-situ methods are based on generating heat within the producing formation by burning part of the oil in place. Air is injected into the reservoir, and the ignition generally is started artificially. A burning front is producted in the crude oil near the well bore, and it moves radially away from the injection well. Combustion gases and steam produced from vaporization of formation fluids move ahead of the burning front toward producing wells. The steam heats the viscous crude oil and displaces it from the steamed region. Any remaining oil is vaporized as the burning front approaches, producing light gases and a coke-like residue that burns. The gases flow ahead of the burning front and dissolve in the oil. All the hyrdocarbon recovery is from the producing wells. A variation of the technology involves injecting water along with air or alternately. This provides more steam in the producing zone, resulting in improved thermal efficiency. In-situ combustion, which was thought to have great potential, has declined in recent years and now accounts for only 3 per cent of the tertiary-recovery production in the United States.

Injection of a miscible gas into the producing zone can increase oil recovery in several ways, including reduction of capillary forces, vaporization of part of the crude oil, reduction of viscosity, expansion of the oil, and alteration of reservoir rock. The principal gas currently used in this technique is carbon dioxide. Although carbon dioxide is not completely miscible with oil at normal pressures, under high pressure it is highly soluble and causes expansion of the oil and reduction of its viscosity. Carbon dioxide also improves miscibility though extraction of the light hydrocarbons from the oil. It has the additional advantage of high viscosity relative to hydrocarbon gases, which improves sweep efficiency and retards gravity separation (Figure 2.1.2.3B).

Three principal techniques are used in CO_2 injection, depending on the miscibility characteristics of the crude oil: (1) injection of CO_2 alone, (2) injection of CO_2 followed by water or carbonated water, and (3) injection of carbonated water. The amount of CO_2 required for a project depends on the size of the reservoir and its pore volume, but because of the need for large CO_2 supplies, the technique generally is applicable only on a fairly large scale.

Micellar polymer injection is a form of water flooding that involves the sequential injection of surfactants to reduce surface tension in the reservoir sediments followed by water to drive the surfactants ahead toward the producing wells. Although much research had gone into the development of the technique, it has not proven economically viable in most field applications and is little used at present. A variant of the micellar flooding method is the so-called improved water flooding, which employs a high molecular weight additive, generally a polyacrylamide or polysaccharide, in the injection water to increase the viscosity of the displacing water. This has the effect of improving sweep efficiency through increasing the proportion of the producing zone swept by displacing water and reducing the tendency of the water to bypass residual oil. In general, the polymer-injection techniques have not gained wide acceptance, because they give only a relatively small incremental additional recovery above that obtained by conventional water flooding and entail considerable additional expense.

2.1.2.4. Water use. Water used in oil and gas production includes water used for mixing drilling mud, that used in hydraulic fracturing to increase permeability, and, of much greater significance, that used in water flooding and tertiary recovery processes.

Water used in drilling mud can be considered to be consumed. Unsuccessful wells are generally left full of drilling mud, with a cement plug in the upper part of the hole. When a well is placed in production, the drilling mud is removed; however, essentially none of the water content returns to the water cycle, because the mud is either salvaged for reuse or permitted to dry out. This water use, although it aggregates to an impressive volume for the United States as a whole, locally represents a one-time use and has little impact on other water users. Davis and Velikanov (1979) reported that about 46 x 10^6 m^3 of fresh water were used this way annually in the United States.

Larger volumes of water are required for water flooding and tertiary-recovery projects. Water flooding should, over the long term, require about 1 volume of water per volume of oil displaced; however, in some fields there could be an initial requirement to saturate parts of the producing zone that had been drained by gravity in the primary-production phase. A rough estimate of the unit water requirement is 1 to 3 m^3 of water per metric ton of oil produced. Buttermore (1966) reported that as of 1962, the total demand for water for secondary recovery in the United States was 690 x 10^6 m^3 annually, of which about 30 per cent was for fresh water. As about 35 per cent of current United States oil production is from water flooding, the annual water requirement would be close to the 1979 secondary-recovery production of 270 million metric tons, or about 300 x 10^6 m^3 of water.

The water-to-oil ratio for tertiary recovery is more complex. In the steam-soak system, the injected water is later produced with the oil and thus is available for reuse. In the steam drive, miscible-gas injection, and micellar-polymer injection systems, the injected water acts as in a conventional water flood, displacing oil from the producing zone and driving it forward toward the producing wells. Ideally, no water is produced until all the oil is flushed out; however, this condition is never achieved, and the typical water flooding project results in an increasing ratio of water to oil production until the rate of oil recovery drops to an uneconomic level. Thus, the water produced with the oil is available for reinjection, but ultimately a volume of water equivalent to the volume of oil displaced is left in the producing zone.

Data from the USSR (Davis and Velikanov, 1979) indicate a water requirement for drilling, flooding, gathering, and treatment of produced oil of 2.5 to 3.5 m^3 per metric ton of oil produced, much more than United States and Canadian estimates; the term "treatment" may include refining steps as used herein.

Water used in gas production is limited to that required for drilling mud and for hydraulic fracturing, both of which are relatively small one-time uses that have little impact on local water balances.

2.1.3. Oil shale

The use of water for oil-shale production currently is of little consequence in most of the world. Although very large oil-shale resources are available in the United States, Australia, Brazil, and elsewhere, the high economic cost of extracting the oil has discouraged development. As of 1982, the estimated world production was about 10 million tons of oil product per year, as follows: the USSR, 7 million; People's Republic of China, 3 million; and Brazil (demonstration plant), 0.1 million. A single plant under construction in the United States under a government financial guarantee is expected to begin production at a level of 0.5 million metric tons per year in the late 1980s. Because the oil content of even rich shales is only about 10 per cent by weight, it is impracticable to ship the raw shale great distances for procesing. Accordingly, thermal processing is performed on site to produce a crude oil, which is transported by pipeline to conventional refineries for further upgrading. The need for on-site processing has important consequences with regard to water resources because much of the consumptive use of water will occur in water-deficient areas.

Oil shale is a fine-grained sedimentary rock containing varying amounts of organic material termed "kerogen." When heated to 482° C, kerogen decomposes into hydrocarbons and a carbonaceous residue. Upon cooling, the gaseous hydrocarbons condense into a viscous petroleum fluid, which can be refined by conventional processes into a variety of fuels or can be burned directly for electric-power generation.

Commercial production of liquid fuels from oil shale dates from 1838 in France, and oil was first distilled from shale in the United States and Canada in the 1859s. An oil-shale operation was carried on in Scotland as recently as 1962, when it closed down because of inability to compete with petroleum. Some of the richest and most extensive deposits are found in the United States in the Green River Formation in Colorado, Utah, and Wyoming (Figure 2.1.3). Considering only high-grade shale yielding more than 85 liters per ton, the reserves in the Green River Formation are estimated at 82 x 10^9 tons of oil. These rich deposits will see the first exploitation; however, even larger deposits of lower grade shales are available from the Green River Formation, and in shales of Paleozoic age in the eastern United States.

Shale oil can be produced from oil shale in three basic modes: (1) surface retorting, (2) in-situ conversion, and (3) modified in-situ conversion. In surface retorting, the shale is excavated by either surface or underground mining and transported to on-site facilities, where the oil is extracted in the thermal-processing plant. " In-situ retorting" refers to a method in which all heating by combustion in accomplished underground. The modified in-situ method involves mining about 20 per cent of the ore and rubblizing, underground, large volumes of ore which are then fired to drive out the hydrocarbons with an underground burn. Surface retorting has the advantage that the process can be readily monitored and controlled but the disadvantages that large volumes of material must be mined and transported to the retort and large volumes of waste must be disposed of in an environmentally acceptable manner. The in-situ method avoids the problem of disposal of waste shale but results in lower recovery of oil. Also, large-scale burns are difficult to maintain because the shale swells with heating and tends to choke off the burn. The modified in-situ method is a compromise that results in less waste to dispose of than surface retorting and less handling of materials but a lower oil-recovery ratio.

2.1.3.1. Surface retorting. This system begins with mining the shale either in surface mines or by underground methods. Surface mining, of course, is restricted to areas where ore-grade shale occurs close to land surface. In principle, surface mining should be straight-forward,

Figure 2.1.3. Oil-shale strata exposed along Colorado River cliffs near Rifle, Colorado, USA. The darker, more-resistant beds are the richest in kerogen content. (Photo by U.S. Bureau of Land Mangement.)

with removal and stockpiling of modest amounts of overburden followed by excavation of the oil shale with conventional equipment by methods similar to coal mining. The amounts of material to be moved in a commercial-scale operation would dwarf that of most coal mines, however. For example, a commercial-scale retort of 13,600 metric tons per day capacity would require input of 90,000 metric tons per day of rich shale with 135 liters per ton oil content, which is very high grade shale. Such a plant would require 33 million metric tons per year input, far more than the output of the largest coal mines. Because the shale bulks during the retorting process, an even larger volume of waste product must be disposed of.

Underground mining is the preferred technique at sites where the oil shale crops out along cliffs, as, for example, along the Colorado River in the United States. Canyons in some areas provide ready access so that rich ore zones can be mined-out selectively. In favorable locations such as these, underground mining offers several advantages over surface mining. Most of the rich seams are relatively thick, on the order of 20 m, and have few faults and fissures. The ore is dense, consolidated marlstone that resists compression and shear stresses, and because they are above the local water table, the workings are quite dry. The accepted mining method is room-and-pillar mining, with the ceiling support provided by roof bolting. Rooms are about 20 m square, with pillars of comparable size. Because of the great thickness of the ore, it is mined in two levels (Figure 2.1.3.1A).

In either method of mining, the ore is transported to an on-site retort for thermal processing. First the ore is crushed and sized, and then it is retorted in either a direct or an indirect heating facility. The principal indirect method involves transferring heat to the ore by ceramic balls. Crushed shale preheated to 260°C is mixed with ceramic balls heated to 650°C in a pyrolysis drum. As the drum rotates, the shale temperature is raised to 482°C. The shale-oil vapors are collected, cooled, and fractionated. The spent shale and ceramic balls are separated, and the retorted shale passes through a heat exchanger to produce steam for plant use. The shale is then quenched in water and the moisture content is adjusted to a suitable level for disposal; the ceramic balls are cleaned and recycled.

In the principal direct-heating method, a refractory-lined vertical kiln is employed. A moving bed of sized-shale granules, flowing downward through the kiln, comes in contact with a counterflow of hot gases, which pyrolyze the shale's organic constituents and convey the resulting vapors out the top of the retort. In this process, the residual carbon in the retorted shale can be used as a source of heat for the retort, or the retorting heat can be provided by recycling the product gas, heated in the external furnace. Wastes from either method can, to a degree, be placed in worked-out areas; however, the material swells on heating, so that the volume of waste exceeds the available worked-out area and some waste must be placed on the surface (Figure 2.1.3.1B).

2.1.3.2. In-situ conversion. This method is most applicable in shallow deposits (less than 300 m depth), because it is necessary to create porosity for the burn to proceed. This is done by hydraulic or explosive fracturing. The shale is then ignited at a well bore and produced vapors are driven forward ahead of the burn front to production wells, where they are collected. Despite considerable research on in-situ technology, it does not yet appear to be commercially practicable on a broad scale.

2.1.3.3. Modified in-situ conversion. The modified in-situ methods provide for the porosity required for burning by mining out about 20 per cent of the shale beneath a rectangular area. Vertical shot holes are drilled through a large mass of the shale, which is then shattered with explosives, producing a silo-shaped mass of rubble. This underground prism of shale is then ignited at the top, and the burn proceeds downward, driving ahead gases and liquids which are drained off at the bottom. The flame front is regulated from the surface by injection of steam and air or oxygen. The reason for employing this hybrid form of underground mining followed by burning in place is that oil shale expands upon retorting and the expanded shale would tend to choke off the burn if provision were not made for the bulking effect. The planned scale for underground retorting is grand. On Federal lease C-a, in Colorado, U.S.A. for example, the planned full-scale chambers will be 20 x 50 m in plan and 133 m high. There are a number of engineering alternatives for achieving uniform rubblization and burning, but all follow the basic scheme described above (Figure 2.1.3.3).

The modified in-situ method entails major changes in the ground-water regime over the area of development. Because the retorts must be dewatered to permit burning, pumping over broad areas is required. To the extent that the oil shale forms an aquifer, the aquifer is destroyed within the area of combustion, and the large masses of burnt shale left in place when resaturated could leach noxious solutes for the indefinite future. Because most of the commercially attractive oil-shale deposits in the United States are in remote, semi-arid grazing areas and ground water is little used, the disruption of aquifers due to the modified in-situ method is not expected to be of great economic import.

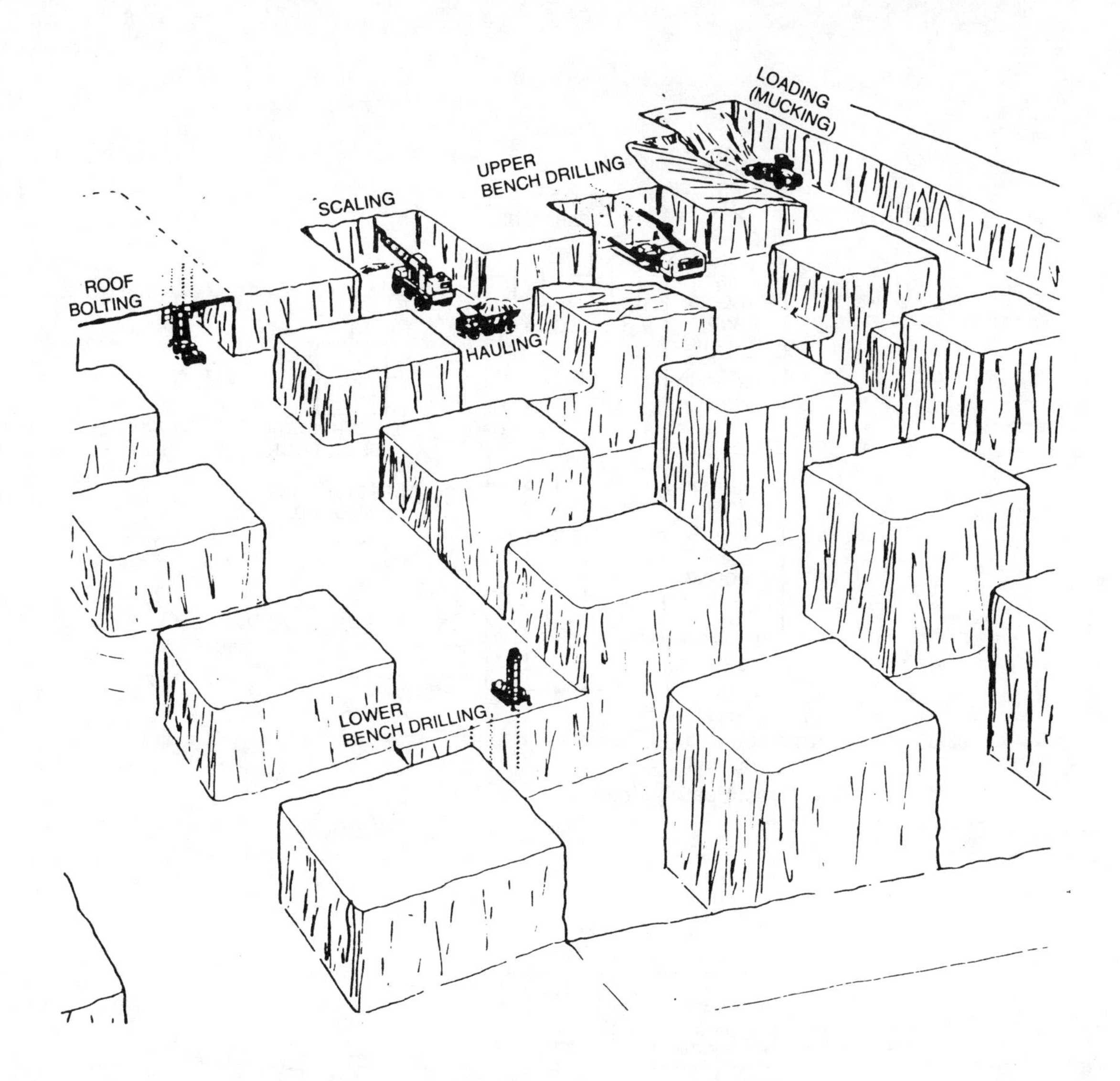

Figure 2.1.3.1A. Underground mining of thick seam of oil shale by the room-and-pillar method. (Source: U.S. Office of Technology Assessment, 1980.)

2.1.3.4. Water use. Water plays an important role in all aspects of oil shale production. Mining requires water control, if not dewatering; in-situ processes require dewatering for the burn to proceed and large volumes of steam to control the burn; surface retorts require water for cooling and quenching; accessory electric power plants require water for cooling; and disposal of spent shale requires large amounts of water.

Early plans for oil-shale development employing surface retorting (U.S. Department of the Interior, 1973) estimated water requirements for a plant with a 5×10^6 metric tons per year output at an average 19×10^6 m^3 per year of water, or almost 4 volumes of water per volume of shale oil. Almost half of this water demand is accounted for in the waste disposal phase. Recent estimates of water demand for the modified in-situ methods are much less, about 1 volume of water per volume of shale oil output (Davis and Kilpatrick, 1981). The lower demand is due mainly to (1) reduced demand for processed shale disposal due to the fact that only 20 per cent of the shale requires surface disposal, coupled with the fact that present operating plans are to moisturize the shale to 15 per cent water content by weight compared with 20 to 30 per cent in early plans, and (2) using gas turbines in lieu of steam-electric plants for accessory power production.

A novel aspect of the modified in-situ plan for Federal lease C-a is that dewatering required for operation of the underground retorts is expected to satisfy the consumptive demand and leave a surplus of water that will be disposed of by off-site reinjection to the ground-water body. Under the plan, no external source of water would be required for the

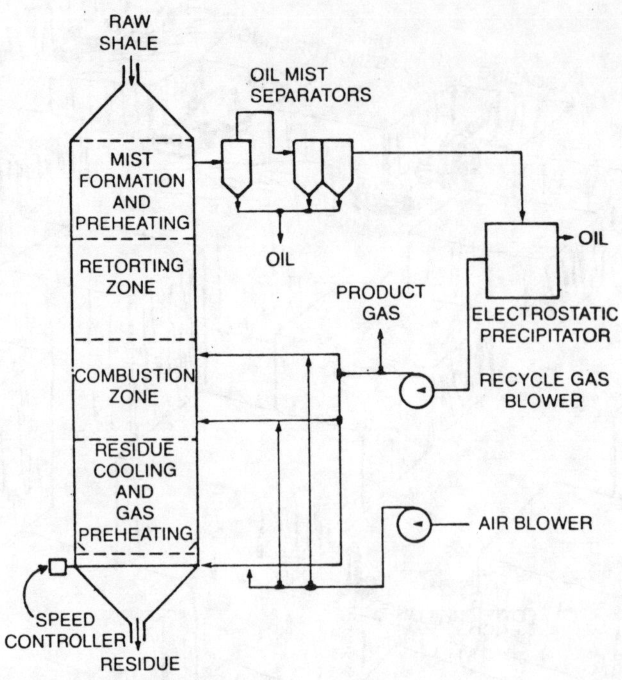

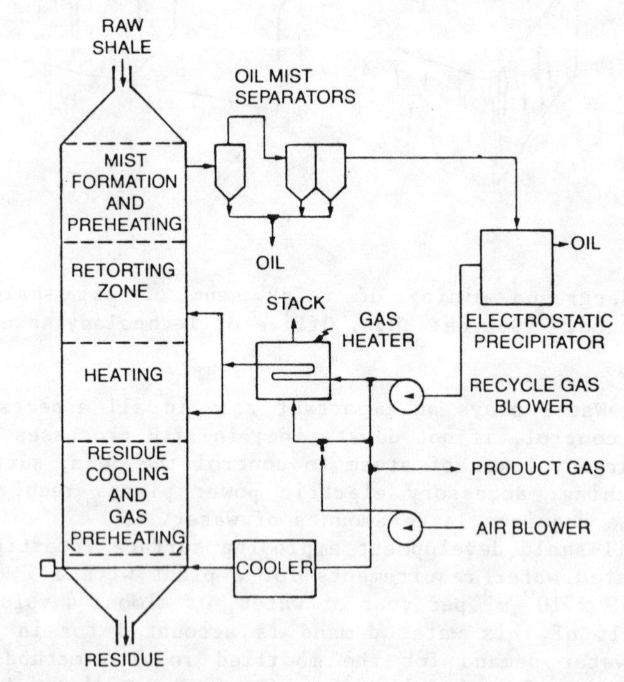

Figure 2.1.3.1B: Vertical-kiln type of surface oil-shale retort: (a) Direct heating mode, (b) Indirect heating mode. (Source: U.S. Office of Technology Assessment, 1980.)

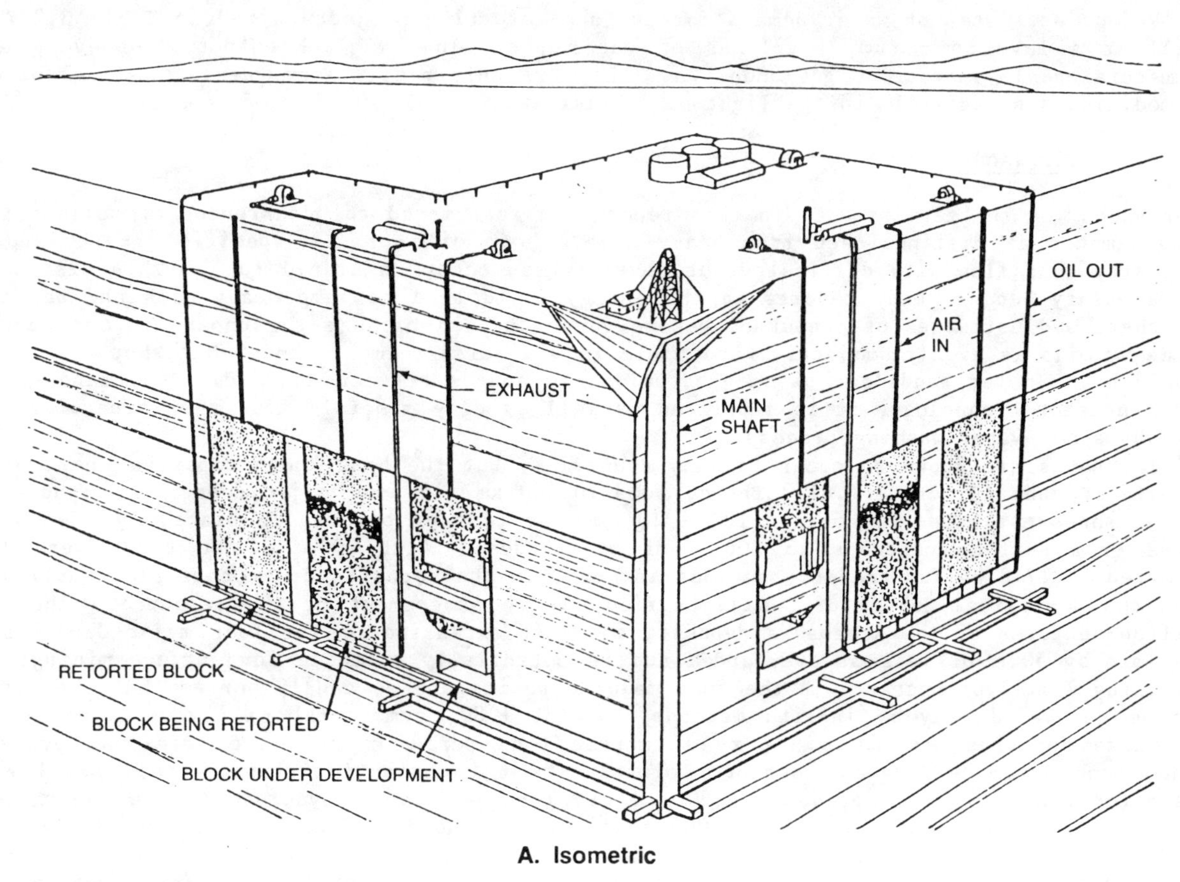

A. Isometric

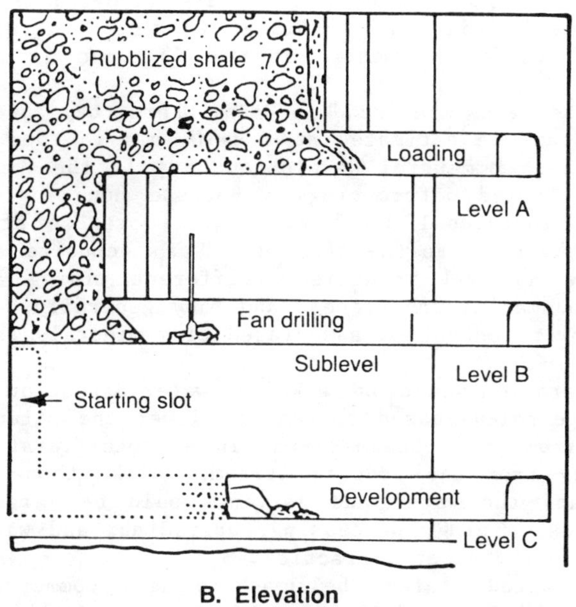

B. Elevation

Development proceeds simultaneously on several levels. At each level, horizontal drifts are driven the full width of the retort block, and a vertical slot is bored to provide void volume for blasting. About 20% of the broken shale is removed after each blasting operation. The rest is left in the retort volume.

Figure 2.1.3.3. Modified in-situ method of oil-shale development. (Source: U.S. Office of Technology Assessment, 1980.)

operation. Nonetheless, the consumptive demand, although met by on-site sources and producing a surplus for disposal, is still a real consumptive use. The water consumed represents water withdrawn from ground-water storage, which ultimately would be replenished by lateral subsurface inflow after the subsurface retorting and dewatering operations are completed.

Recent estimates of water demand for surface retorting projects cited in Table 3.2 A are still at a level of about 3 volumes of water per volume of oil output. However, until commercial-scale operations get under way, they are only preliminary guesses and may have to be modified substantially in the light of experience.

2.1.4. Tar sands

"Tar sands" as the term is used in this report are restricted to occurrences of solid bitumen in sediments as distinguished from heavy oils, those oils of low specific gravity and low viscosity that flow with difficulty but nevertheless occur in liquid form. This distinction is necessary because many reports in the energy field tend to lump heavy oils and tar sands together, and estimates of resources become very confused because the upper limit of density of heavy oils is highly subjective and there is wide disagreement about what that upper limit should be. In tar sands the bitumen occupies the spaces between the individual sand grains. Under present technology, it is separated as well as clay minerals and other mineral matter, only by a hot-water washing process.

Tar sands are known to occur in many countries, but the best known deposits are those in Alberta, Canada (Figure 2.1.4). There, some 70,000 km^2 in four major deposits are underlain by tar sands estimated to contain 122×10^9 metric tons of bitumen in place. In depth they range from surface outcrops to 800 m of overburden. Although the tar sands were first reported in 1778 by fur traders, serious interest in the deposits did not develop until 1897, when the Geological Survey of Canada carried out an assessment of the resource. About 0.3 million hectares of the Athabaska deposit, the most extensive of the four major deposits, is overlain by 50 m or less of overburden and is potentially suitable for surface mining. The remaining 7 million hectares in the four major deposits are so deeply buried that the bitumen can be recovered only by in-situ methods. As of 1980, commercial oil production from the Athabaska deposits was about 6.2×10^6 metric tons per year from two prinicpal projects, Suncor and Syncrude Canada. Suncor (formerly Great Canadian Oil Sands, Ltd.) has been in operation since 1967 and produces 6,800 metric tons per day. Syncrude Canada, which began operation in 1978, produces about 10,000 tons per day and is expected to reach full capacity of 17,000 tons per day in 1984. These two operating projects account for some 0.3×10^9 metric tons of reserves. In addition, 5×10^9 metric tons of bitumen are recoverable by surface mining at other sites, and 10×10^9 metric tons are estimated to be potentially recoverable by in-situ methods. An additional 84×10^9 metric tons of bitumen is not recoverable by current technology.

In the United States, the principal tar-sand deposits are found in Utah, where they appear to represent exhumed oil fields. Resources in Utah are estimated at 4×10^9 metric tons of bitumen in place. The deposits most likely to see commercial development are in the Uinta Basin, where the tar sands occur in several scattered outcroppings. Because none of the deposits are of the scale of those in Alberta, individual developments are foreseen at a smaller scale than the massive operations in Alberta. Furthermore, the Utah deposits are physically different from the Alberta tar sands and will require a different mining and separation process. The Utah deposits occur in nonmarine sandstones, and they generally have lower bitumen content and higher viscosity but lower content of associated clay minerals than the Alberta tar sands.

The leading separation processes under consideration would use alkaline water digestion at ambient temperature but with the addition of a petroleum-based solvent to lower the bitumen viscosity. The bitumen would then be separated from the process stream in a froth-flotation step. The sand tailings, which separate readily from the process stream, unlike those of Alberta, would be routed to a disposal pile, and the water and solvent would be largely recycled. With the exception of these differences, mining, processing, upgrading, and water consumption in the Utah deposits will be similar to the Alberta practice.

Another, somewhat different deposit in the United States, believed to have commercial potential, is an oil-saturated diatomite found at McKittrick in the San Joaquin Valley of California. This deposit, which underlies about 1,000 hectares, is believed to contain 113×10^6 metric tons of oil. Present plans of the Getty Oil Corporation call for extraction employing surface mining and a surface extraction plant at a rate of 2,700 metric tons per day over a period of 48 years. Diatomite would be produced as a byproduct (Morton, 1981).

Figure 2.1.4. Bucket-wheel excavator at tar sand face, Suncor mine, Alberta, Canada. (Photo courtesy of American Petroleum Institute.)

In the USSR, a tar-sand deposit at Yarega in Siberia has been worked for several years by an underground mining operation involving steam extraction. Little other information on the project is available, but the cumulative production is reported to be 3 x 10 metric tons.

2.1.4.1. Mining and extraction. The longest experience in tar-sands production has been at the Suncor project in Canada. Suncor provides the background for this description. Initially, the severe winter climate presented severe problems. Equipment tended to freeze in the arctic temperatures. The tar sand froze to the consistency of concrete and ripped the teeth out of bucket-wheel excavators. With experience, however, these problems were overcome, and the project has been operating profitably at a steady rate for several years.

The tar-sand deposit crops out along the east-facing bluff of the Athabaska River about 80 m above river level. A sharp escarpment, which forms the eastern boundary of the tar-sand deposit, drops to a gently sloping area bordering the river where the processing facilities are located. The surface of the mining area is relatively flat and is covered with muskeg, small trees, and brush. The undulating surface of the underlying limestone beneath the tar sands stands about 7 m above the river. The thickness of overburden and tar sand, the ratio of overburden to tar sand, the bitumen saturation, and other properties of the deposits are highly variable.

Within the pit limits, which encompass about 90 per cent of the lease, the overburden averages 17 m thick, ranging from 0 to 30 m. The economic tar sands average 43 m in thickness, ranging from 0 to 78 m. The bitumen content of the economic tar sands ranges from 8 to 18 per cent and averages 12 per cent by weight on a dry basis. The material smaller than sand size ranges from 5 to 45 per cent by weight.

The material making up the overburden from the surface down consists of muskeg, glacial drift, the Clearwater Formation of early Cretaceous age, and a zone of uneconomic tar sand in the upper part of the early Cretaceous McMurray Formation, containing less than 8 per cent bitumen. The overburden is the source of most of the material used in construction of earthworks in the pit. The amount of stripping required ranges from 7 to 23 x 10^6 metric tons per year.

The tar-sand deposit proper consists of a basal fluviatile sand overlain by lagoonal deposits, which have been eroded and refilled with silts, some of which are now cemented with siderite. The bitumen content is highly variable, ranging from 0 to 18 per cent by weight. The deposit also contains numerous thin clay layers which are barren. In general, the grain size and bitumen content increase toward the base of the McMurray Formation.

The mining process includes land clearing, stripping of muskeg, overburden removal, construction of retaining dikes, tar-sand excavation, and extraction of the bitumen in the processing plant. Because the muskeg is very wet, an extensive network of ditches followed by 2 years of drainage is required before other operations can proceed. The muskeg is then stripped and kept within retaining dikes up to 30 m high. Most of the rest of the overburden is used to build retaining dikes for future tailings. These structures are up to 100 m high, are constructed with surface slopes of 1 to 2.5 per cent, and have clay cores.

The average thickness of the tar sand is about 50 m. Mining is done with two bucket-wheel excavators operating on separate benches, one above the other. The tar sand is transported to the processing plant by continuous conveyer.

The fundamental steps in extraction are feed conditioning, separation of bitumen, waste disposal, and cleaning the bitumen concentrate. Conditioning is accomplished by mixing the tar sand with water and caustic soda at 162°C to bring the pH up to 8.0 to 8.5. This is done in a rotating horizontal drum 6 m in diameter by 17 m long. Tar sand, hot water, and soda are injected into the feed end of the drum, and steam is injected under the pulp formed to maintain the temperature in the range 162°-172°C. The conditioned pulp is discharged to a feed sump and then is introduced into separation cells, vertical cone-bottom vessels 15 m in diameter and 8 m high, with a center feed well into which the pulp is pumped. The froth that floats is skimmed by rakes to a peripheral collector and then is pumped to the final extraction plant. The coarse mineral matter, which settles to the bottom cone, is removed and pumped to the tailings pond (Figure 2.1.4.1).

Froths arriving at the final extraction at 142°C are heated with steam and diluted with naphtha to reduce the viscosity. This blend is then separated in a conventional centrifuge process which removes any remaining mineral matter. The diluent and water are separated by distillation and the diluent is recycled. The bitumen is then introduced into a coking drum and heated to about 480°C. The bitumen breaks down into light hydrocarbons and heavy coke, which is deposited in the drum. The hydrocarbons pass to a fractionating tower, where they are separated into light gases, naphtha, kerosene, and gas oil. The light gases and gas oil are used as fuel in the plant and mining equipment, and the naphtha and kerosene are shipped by pipeline to a standard refinery, where they are converted to saleable products.

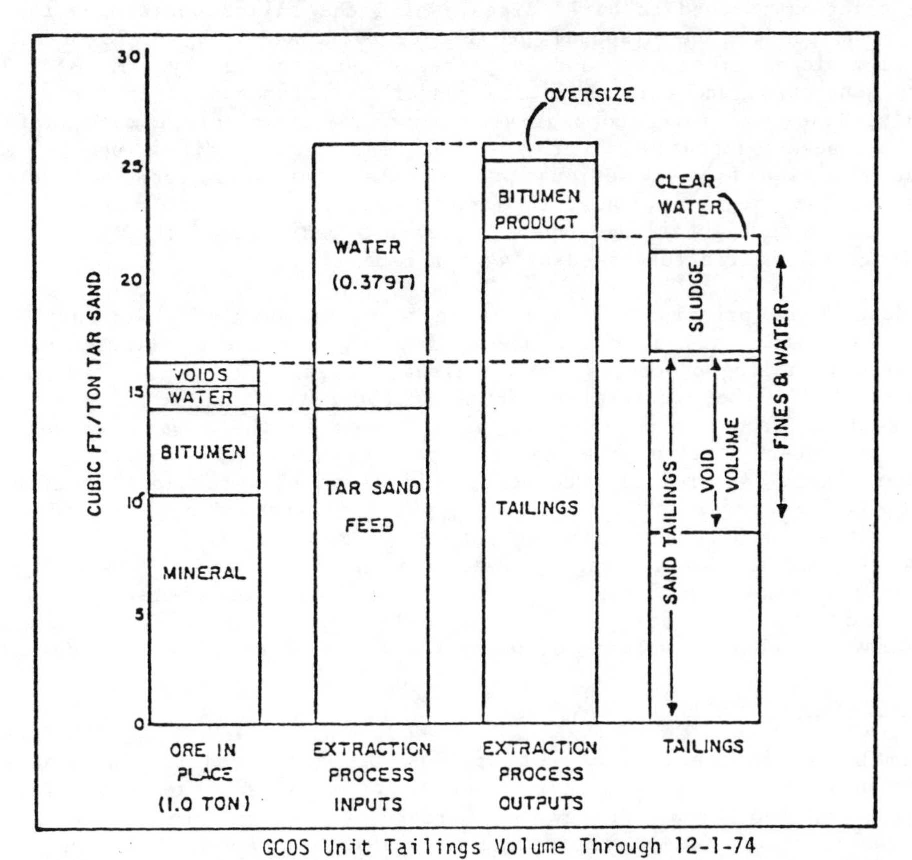

GCOS Unit Tailings Volume Through 12-1-74

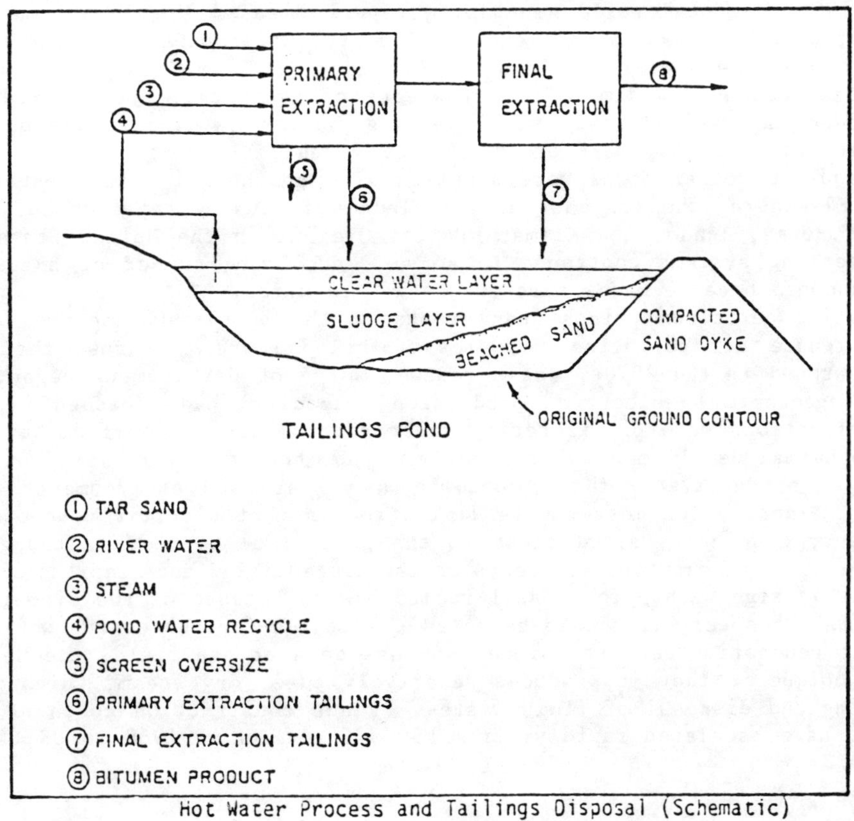

Hot Water Process and Tailings Disposal (Schematic)

Figure 2.1.4.1. Process streams and tailings disposal at Alberta, Canada, tar-sands project. (Source: Camp, 1976.)

The tailings stream, about 1,500 liters per second, is delivered to the tailings pond. Here the coarse fraction is used to build dikes, while the slimes containing the fines flows to the center of the pond. The disposal problem centers about physical containment of the tailings. When a cubic meter of tar sand is processed, approximately 1 m^3 of solids and 0.3 m^3 of sludge are generated, and this obviously cannot be accommodated in the original space. Because the sludge loses its water content very slowly, the principal method of disposal to date has been permanent impoundment of the material behind dikes up to 100 m high. Recognizing that this would be a serious problem over the long term, the operators have endeavored to reduce the volume of sludge through more selective mining and reducing the water content before delivery to the tailings ponds. Over a 5-year period they were able to reduce the volume of sludge per metric ton mined by 44 per cent.

2.1.4.2. Water use. The principal water uses in tar-sand mining and extraction are in the fluid process train, which ends as a component of sludge in the tailings pond, and makeup water for the various cooling processes of the plant, chiefly the 76-MW steam-electric power plant which generates all the electricity for the 6,800 metric ton per day Suncor facility. Much of the waste heat from the power plant is salvaged in the separation process, so the consumptive use of water attributable to electric generation is much lower than in a conventional power plant. Accordingly, the consumptive use of water in the entire process is represented by the water entrained in the sludge, which is trapped over the short term, plus miscellaneous cooling requirements throughout the process for which information is unavailable. A diagram of the tailings stream presented by Camp (1976) (Figure 2.1.4.1) indicates a water requirement of 0.379 ton of water used per ton of tar sand mined. At an average bitumen content of 12 per cent, this would suggest a water requirement, almost entirely consumptive, of about 3 volumes of water for each volume of product output.

2.1.5. Geothermal development

Although geothermal heat in the form of hot springs has been used by man since prehistoric times, it did not become a significant energy source until the 20th century. Electricity was first generated using geothermal energy at Larderello, Italy, in 1904, and the first commercial-size generator of 250 kW was installed in 1913. However, the pace of development was slow, and by 1969 the total world generating capacity was only 678 MW, mainly in three fields--Larderello; Wairaki, New Zealand; and the Geysers, California, USA. At that time, the geothermal operations at the Geysers Field were only marginally profitable. A major offshore gas discovery in New Zealand had dimmed the prospects of geothermal energy production there, and Larderello production was declining slowly because of reservoir-pressure decline. Aside from these electric developments, most of the use of geothermal energy was in Iceland, where the capital city and many other towns were supplied with geothermal space heating by central utility systems. Elsewhere, the low cost of petroleum was causing retrenchment of geothermal development, as at Boise, Idaho, and Klamath Falls, Oregon, in the United States, where old geothermal space-heating systems continued to serve a declining number of homes and offices but were being abandoned because major maintenance was required.

Rapidly escalating petroleum prices and threats to the supply of imported oil, together with pressure to reduce air pollution from fossil-fueled sources, turned the economics of geothermal energy around in the 1970s, and a renewed surge of development began. By January 1981, world-wide geothermal electric production capacity had reached 2,300 MW and direct-heating applications were proliferating under governments' encouragement. Much of the enthusiasm for geothermal development of the early 1970s, however, was based on ignorance and a simplistic lack of appreciation of the economic, physical, and environmental handicaps of geothermal energy. Probably the most severe limitation is that electric development requires small, low-efficiency generators sited close to the geothermal source. Although no fuel is consumed, development and maintenance costs are substantial, and capital investment in turbines is relatively higher than for fossil-fueled plants because of the dispersed nature of the resource. In this respect, it should be noted that three-fourths of the world increase in geothermal electric generating capacity since 1969 has been at one field, the Geysers (Figure 2.1.5A), which is unique in that it produces relatively pure, dry steam, which minimizes the problems of handling and disposal of fluid wastes. Plant construction costs at the Geysers, as in most places, have escalated rapidly, from US $140 per KW in 1968 to US $410 per KW in 1980 (Kestin, 1980).

Figure 2.1.5A. Geysers Geothermal Field, Calfiornia, USA, during multiple-well production test. Under normal operating conditions little steam is released at wells. Most steam release is from power plant cooling towers, left horizon. (Photo by U.S. Geological Survey.)

Development of geothermal energy has much appeal in the current climate because it appears to offer the possibility of cheap, clean electric power development. However, the following are essential for successful economic use:

1. Source of heat. The most favorable areas are those of natural high heat flow, the tectonically active belts of the world, such as the "ring of fire" surrounding the Pacific Ocean Basin (Figure 2.1.5B).

2. A trapping structure, at depth reachable by wells, that confines steam and/or hot water in permeable rocks. The permeability may be intergranular or due to fractures (Figure 2.1.5C).

3. A demand close to the source of the geothermal energy for electric power, or for space heating, or other use of low-grade heat. Proximity to the source of heat is most important because steam can be moved no more than about 10 km economically without excessive heat loss in transit.

4. A suitable means of disposal of noxious wastes, chiefly brines but commonly containing relatively high concentrations of toxic elements, such as arsenic, selenium, mercury, boron, and radon.

Geothermal resource development requires major capital investment. The major costs are for well drilling, construction and operation of steam or water lines, and construction and maintenance of electric power plants and transmission facilities. Exploration for and production of geothermal fluids is similar to petroleum operations and requires relatively few but highly skilled personnel having many of the skills required in the petroleum industry. Power generation is very like conventional steam-electric plant operation and has similar low personnel requirements. The major difference is the elimination of the fossil-fuel cycle, which in the case of coal requires a large work force.

Geothermal reservoirs can be broadly classified as (1) vapor-dominated systems producing mainly steam or (2) water-dominated systems producing hot water that contains a variety of minerals. The vapor-dominated systems have the most economic appeal because the steam produced by wells is relatively free of dissolved constitutents and can be fed directly to turbines, and the residue can be used for power-plant cooling requirements or be disposed of readily by reinjection to the producing zone. Unfortunately, there are only two sizable vapor-dominated systems in the world--the Geysers in California, USA, and Larderello in Italy.

The water-dominated systems, such as those of New Zealand, Iceland, the Imperial Valley in California, and the adjoining Cerro Priesto area in Mexico, pose serious engineering and economic problems in operation of power facilities owing to problems with scaling, saline waters, corrosion, and disposal of waste fluid and mineral matter. The environmental problems include air pollution by accessory noncondensible gases, leakage of brine, land subsidence stemming from production of large volumes of water, and the hazard of triggering earthquakes through water withdrawal or reinjection.

In both systems, a depletable resource--heat contained in water--is being mined, and the lifetime of the project depends on the amount of heat stored in the given reservoir and the presence of water to serve as a heat-exchange medium. The natural, equilibrium heat flow, even in areas of high natural heat flow, is not sufficient to support major long-term energy production under present-day economics and technology.

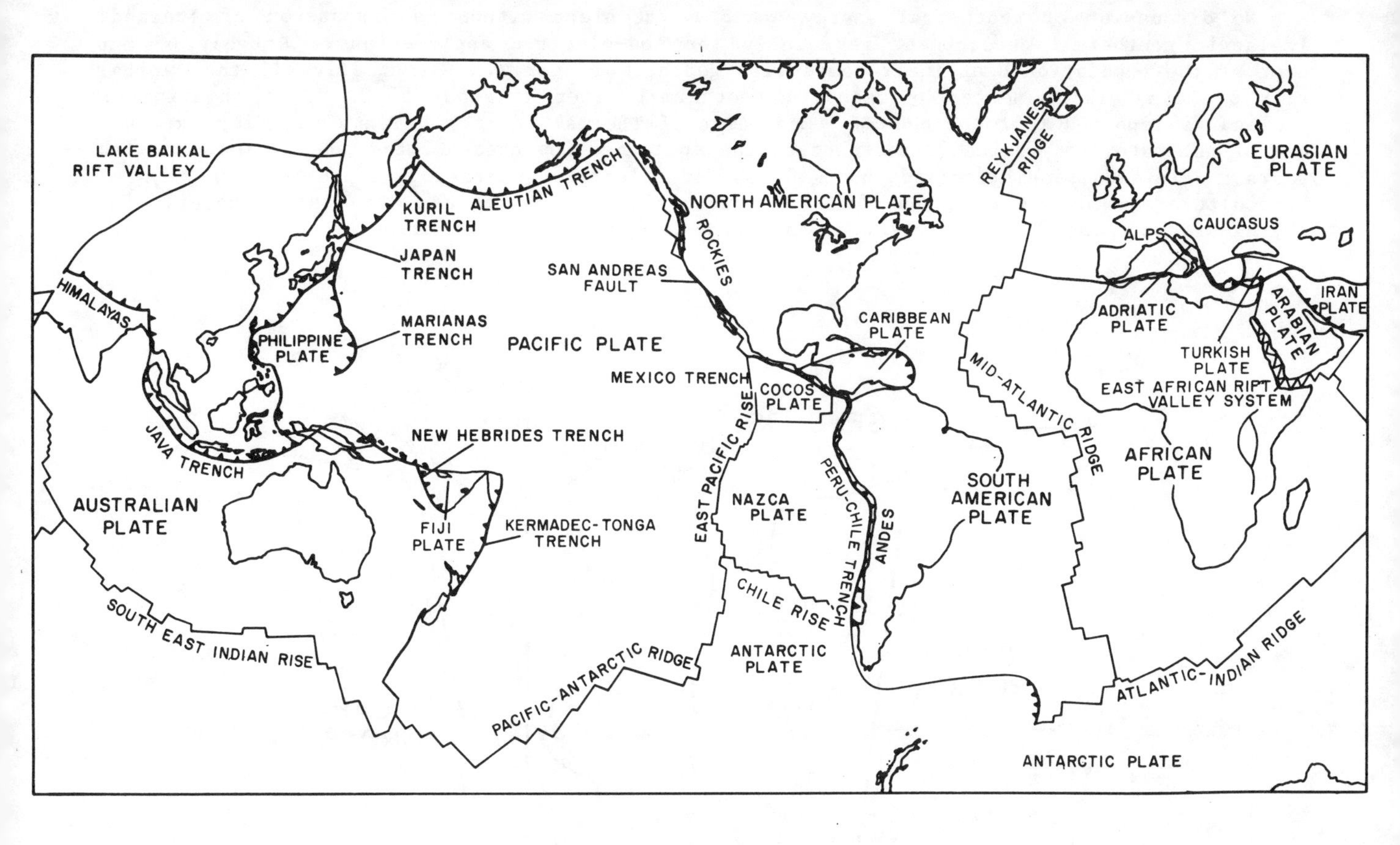

Figure 2.1.5B. Map of tectonic plate boundaries, the sites of most tectonic activity, volcanism, and abnormally high heat flows. Nearly all significant geothermal exploitation is on or near these plate boundaries.

Depletion of the geothermal resource during full operation of a field is manifested by reduction in the reservoir pressure. For example, the reservoir at Wairakei, New Zealand, which originally produced at pressures ranging from 6.3 to 8.8 kg/cm^2, declined by 38 per cent from the beginning of commercial power production in 1959 through 1975 (DiPippo, 1980, p. 213). Such decline of pressure is compensated for in most fields by drilling of additional wells to deliver the design input of steam to the power plant. The long-term pressure decline is a complex function of heat and fluid flow; however, most generating plants are planned for about a 40-year life until the resources is depleted.

The most favorable prospect for geothermal development lies in the discovery of additional vapor-dominated systems in areas of substantial electric power demand. Under present-day electric-power technology, only the larger, high-temperature reservoirs of the water-dominated type appear to offer much promise of development, but many of these are hampered by environmental problems. Geothermal development has its greatest appeal in nations lacking domestic energy resources, such as Iceland, or nations that depend on imports for the bulk of their energy supplies, such as Japan and Italy. Also, in areas where geothermal energy can be tapped readily, it can be a significant replacement for imported energy in space heating, greenhouse agriculture, and fish farming and for industrial processes requiring mainly low-grade heat.

With escalation of petroleum prices since 1970, geothermal development has had strong encouragement. In the United States at the Geysers, steady expansion continued to a producing capacity of 1,020 MW as of January 1983, and hot-water fields have been explored actively for their geothermal potential, although none of these developments are at full commercial scale as of 1982. Japan, the leading producer of geothermal turbines, had reached a capacity of 220 MW as of 1980; Mexico had reached 150 MW, and the Philippines, 279 MW. Neither Italy nor New Zealand has added significantly to geothermal generating capacity, but New Zealand is planning the construction of an additional 150 MW near Broadlands, with the first 50-MW increment due to come on line in 1983 (Kestin, 1980).

No discussion of geothermal energy would be complete without a discussion of Iceland. Icelandic geothermal development is chiefly for non-electric applications. Roughly 65 per cent of the population heat their homes with geothermal energy. Essentially all the capital city of Reykjavik is heated by means of geothermal water at about 86°C. Space heating in Iceland is reported to total 600 MW_t (megawatts of thermal energy). Plans call for expanded use of hot water to provide 80 per cent of the space-heating needs of the country within a few years. In addition, Iceland has 30 MW of electric-generating capacity, 40 MW_t of agriculture/aquaculture use, and 50 MW_t of industrial processing use, the latter chiefly for drying diatomite and for production of salt from sea water by evaporation.

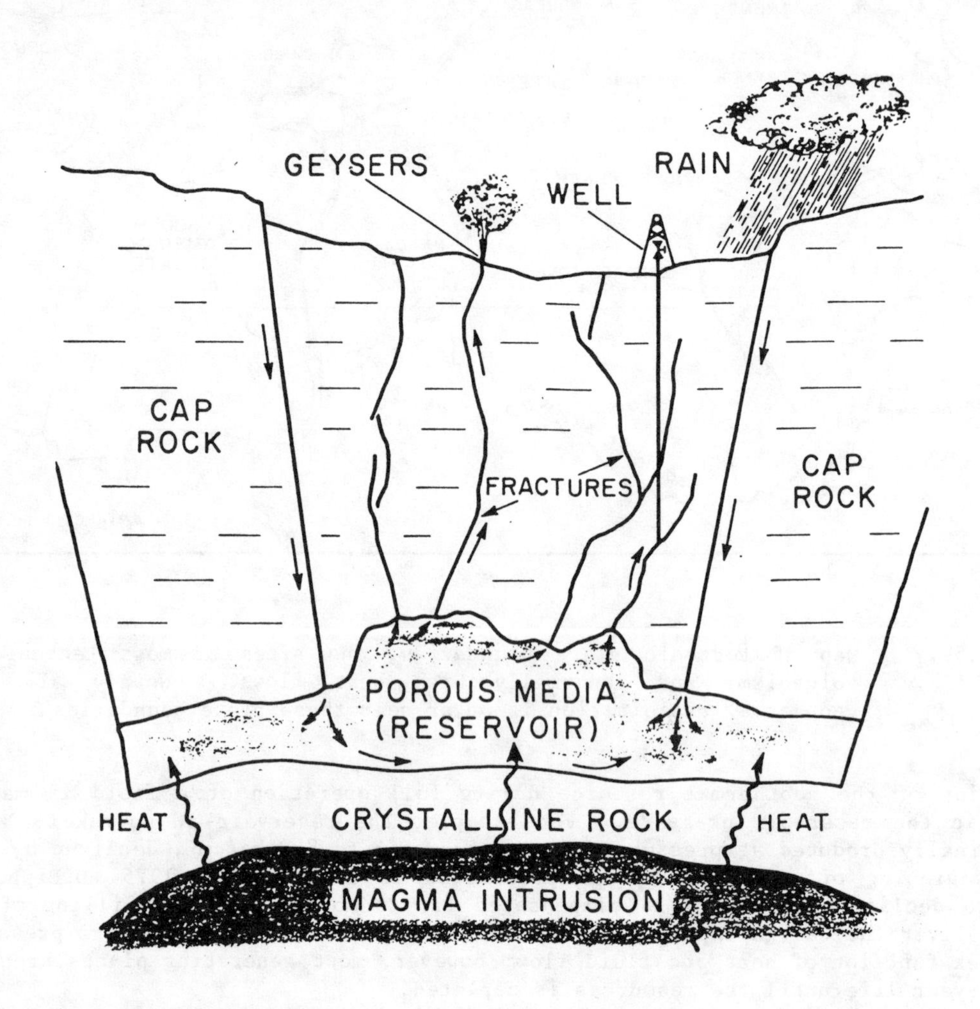

Figure 2.1.5C. Schematic diagram of geothermal system showing major elements: intrusion, reservoir, impermeable cap rock, fractures, and recharge mechanism. (Source: DiPippo, 1980.)

Other leading countries in non-electric applications are Hungary and the USSR. Hungary is reported to use 300 MW_t for space heating, and 370 MW_t for agriculture/aquaculture. The USSR reports using 120 MW_t for space heating and an astounding 5,100 MW_t for agriculture/aquaculture.

2.1.5.1. Water use. In most geothermal systems, essentially no water is consumed in the extraction of the resource as such; however, in electric-power generation a large amount of water is consumed for condenser cooling, as will be discussed under conversion processes in Section 2.4.1. In non-electric applications, the heat is extracted by a variety of heat exchange-mechanisms and the water commonly is returned to the source without significant consumptive use or is simply disposed of to local streams.

At the Geysers (Figure 2.1.5.1), 16 generating units totaling 1,020 MW capacity were in operation as of 1982, supplied by more than 250 wells ranging in depth from 500 to 3,000 m

Figure 2.1.5.1. Geysers Geothermal Field, California, USA, showing production facilities. Wells marked by small steam plumes, center foreground and left center; operating power plant with steam rising from forced-draft cooling towers, left center; and plant not operating, right center. Note large-diameter insulated surface pipelines that carry steam to power plants. Heat loss despite insulated lines requires that power plants be no farther than about 1 km from wells. (Photo courtesy American Petroleum Institute.)

(DiPippo, 1980). A 110-MW unit, the largest size employed, requires about 818 metric tons per hour of steam at 180°C and 8.1 kg/cm^2. About 15 wells are required to support a 110-MW unit; individual wells produce 34-159 metric tons per hour at a wellhead pressure of 9.8 kg/cm^2. To minimize loss of thermal energy in the steam gathering system, insulated pipe 254-914 mm in diameter is used and the plants are spaced so that pipelines are no longer than 2 km. Nonetheless, the thermal efficiency of the best units at the Geysers Field is only about 15 per cent. In water-dominated systems, the highest thermal efficiencies are in the neighborhood of 10 per cent and in some systems are as low as 5 per cent. This inherently low efficiency, due to low temperature and pressure of the natural reservoirs, results in comparably greater water consumption to dispose of the 85-95 per cent of the thermal energy that must be dissipated by evaporation.

2.1.6. Uranium mining and milling

The principal source of nuclear energy is uranium. In a nuclear power plant, the fuel in the reactor undergoes spontaneous nuclear fission; that is, individual atoms in the fuel are split into radioactive fragments, and energy is produced in the form of heat. This heat is used in boilers to convert water into steam for driving turbines to produce electricity. Most of the commercial power reactors in the world are of the light-water type, which technique was pioneered in the United States. They are termed "light-water reactors" because ordinary water, as distinguished from deuteriated or heavy water, is used to moderate the nuclear fission and as a heat-transfer medium. The other principal types of power reactors are the heavy-water type, developed in Canada, which employs unenriched uranium as fuel and deuteriated water as moderator and heat-transfer medium, and high-temperature, gas-cooled reactors which employ carbon as the moderator and hydrogen gas as the heat-transfer medium. These are widely used in the United Kingdom.

The use of nuclear energy as a source of heat dates to the development of nuclear weapons in the United States during World War II. The reactors used for making weapons-grade material emitted great quantities of waste heat, and generation of electric power was an obvious by-product. The first commercial-scale electric-power plant, the Dresden Plant near Chicago, Illinois, was put into operation in 1960. Growth of nuclear electric generation was rapid during the 1960s and early 1970s in the United States and elsewhere, but has slowed notably during the late 1970s owing to high capital-investment cost, environmental and safety concerns, and uncertainty regarding future economics.

Despite slowing of nuclear power growth rates, nuclear electric generation world-wide reached 673 x 10^3 GWh in 1979, about 8 per cent of total world electric generation. In the United States, nuclear generation accounted for over 11 per cent of the total electricity generated in 1980, or 266 GWh. Other leading nations in nuclear electric generation as of 1980 are: France (61 GWh), Federal Republic of Germany (44 GWh), Canada (40 GWh), the United Kingdom (37 GWh), Sweden (27 GWh), and Switzerland (14 GWh) (United Nations, 1981).

The fuel used in most nuclear reactors is uranium. The most common type of reactor, the light-water type, uses enriched uranium--uranium that has been processed to produce a blend of U^{235} and U^{238} in which the proportion of the radioactive isotope U^{235} (4%) is greater than in uranium as found in nature (0.7% U^{235}). Following separation, the fuel has to be processed into uranium dioxide, fabricated into fuel pellets, and placed in gas-tight metal tubes termed "cladding." The metal tubes (fuel rods) are grouped into clusters termed "fuel elements"; the core of a reactor contains an array of such elements.

As nuclear reactors are currently designed and operated, 1 metric ton of enriched uranium fuel will produce about 200 GWh of electricity. The spent fuel elements, in which only a fraction of the uranium has been consumed, are removed after about 3 years service and replaced with new elements. The remaining uranium, fission products, and transuranic elements formed in the fission process are retained within the fuel and cladding. By a process known as reprocessing, much of the uranium in the fuel elements as well as the transuranic element plutonium can be recovered for reuse in power reactors. In this way, the existing uranium reserves can be greatly prolonged. Also, a type of reactor termed "fast-breeder reactors" produces in the reaction additional fuel, which likewise serves to conserve the world supply of nuclear fuels.

2.1.6.1. Mining. Uranium occurs as one of the less abundant elements in the earth's crust; its average abundance is about 4 parts per million by weight. In order of abundance in the rocks of the crust it ranks 50th, somewhat less abundant than arsenic but more abundant than boron. It is found in two principal modes, as vein deposits in granitic-igneous rocks and metamorphosed sedimentary rocks and as precipitates in sediments derived from crystalline-granitic rocks. Some nine minerals account for most of the world's commercial ore deposits. The principal vein minerals are pitchblende and uraninite, both forms of uranium oxide, which occur as fracture fillings in granitic intrusives and associated metamorphosed sediments. In two major uranium districts, Blind River in Canada and the Rand in South Africa, the uranium mineralization occurs in pre-Cambrian sediments within the arkosic matrix in quartz-pebble conglomerates. Here the ore averages about 0.1 per cent uranium oxide. Uranium is also found in association with gold in Canada and South Africa. The world's principal producers of uranium are the United States, Canada, South Africa, Sweden, Spain, and Czechoslovakia.

Most of the uranium in the United States is found in fluviatile sandstones, chiefly of Mezozoic and Eocene age. The commercial deposits are concentrated in the Colorado Plateau Region and in intermontane basins in Wyoming and South Dakota. These deposits are made up of a number of uranium minerals, which typically fill pores in sandstone or replace organic fragments. They are mainly tabular in form, comprising layers from a few meters to as much as 20 m in thickness, and range in size from a few tons to several hundred thousand tons of uranium oxide. Such deposits appear to represent precipitation from percolating ground waters in a reducing environment. Large, subeconomic uranium resources are known to occur in extensive organic marine shales in the United States and Sweden, representing deposition from sea water of uranium in a marine environment.

Reserves of uranium are highly sensitive to market price; an increase in market price can greatly alter the reserve base, which is typically quoted at a specific price level. Recent discoveries of very large uranium ore bodies in northern Australia and declining projections for future demand have severely depressed the world price, and many mines depending on lean ore have been closed in the United States and elsewhere.

Mining of vein deposits such as those of Canada and South Africa, is carried out by typical underground mining methods of tunneling and stoping. The crystalline country rocks are generally impermeable, and water plays a negligible role in mining. In the sedimentary ore bodies, such as those of the western United States, the situation is quite different. Being fluviatile sands, the ores commonly constitute aquifers and yield water readily when tapped by mining operations. Underground mines require large pumps running continuously to maintain dry workings. This commonly results in large quantities of excess water to be disposed of on the surface. The most widely practiced method is to permit the waste water to evaporate from disposal ponds, as the water is slightly radioactive and for that reason is not acceptable for discharge to nearby streams.

Near-surface deposits, mined by open-pit methods, generally involve lesser water problems, because most of the mines in the United States are in semi-arid areas and the surface mines commonly are above the water table. Mine dewatering is generally limited to disposal of occasional rain-storm runoff, the water being in demand for water supply at on-site milling facilities.

Lean deposits not suitable for surface mining in some places are worked by an in-situ leaching system. This method entails extraction of the uranium by circulating a solution containing sulfuric acid and an oxidant through the ore zone, employing a series of wells. The acid solution, or lixiviant, dissolves and mobilizes the uranium. The uranium-bearing solution is pumped from the ore zone from a number of recovery wells. The injection and recovery wells are drilled in an established pattern developed from experience and on the basis of theoretical studies. The "pregnant" solution is pumped to a central processing plant where the uranium is stripped by standard methods. After removal of the uranium, most of the barren solution is refortified with acid and oxidant and reinjected into the ore body. The rate of pumping from the production wells is kept slightly greater than the rate of injection to avoid escape of the chemicals from the area of operation; accordingly, a small surplus of water is continuously generated, which is disposed of by evaporation or reinjection to the ore body after treatment.

2.1.6.2 Milling. On-site milling of uranium ore is a process for concentrating the uranium content of the natural ore, which generally makes up less than 0.1 per cent by weight of the material mined. Clearly it is advantageous to concentrate the ore prior to shipping, although small mines commonly ship the raw ore to a custom milling facility. Concentration generally is done by a wet acid-leach process. First, a dominantly tetravalent uranium of uraninite and other uranium minerals are oxidized in a two-step process to uranyl ions. In this process, the most commonly used oxidant is sodium chlorate. The second major reaction involves the use of sulfuric acid, resulting in the ionization in solution to sulfate, bisulfate, and hydrogen ion. The sulfate ion forms complexes of uranium. Depending on the makeup of the ore and the quantity of acid used, various complexes of uranyl sulfates form. The uranyl sulfate is converted to U_3O_8 washed, dried, and shipped in that form to refining plants.

The resulting tailings comprise a residue of ore-forming minerals, the silicate matrix of the host rocks, and chemicals used in the wet-extraction process. Of the ore-forming minerals, most of the uranium and vanadium has been removed; however, many radioactive and toxic metals remain, including thorium, radium, arsenic, barium, copper, molybdenum, lead, selenium, zinc, and others. The silicate matrix consists largely of quartz, commonly accompanied by feldspar and clay minerals. The chief chemicals contributed by the extraction are chloride and sulfate ions in solution at high concentration. A chemical disequilibrium exists between the solution and the various solid phases as well as among the components of the solute. As a result, the various liquid and solid components continue to react over a period of many years after emplacement of the tailings before chemical equilibrium is attained. The various precipitation reactions form easily soluble salts of chloride and sulfate with major cations present and commonly with the trace metals.

Because the residual water is unfit for disposal to either streams or ground water, the general approach to its disposal is to permit the waste fluid to evaporate in ponds or reservoirs constructed for this purpose. This evaporation constitutes the main element of water consumption in the mining and milling phase of the nuclear fuel cycle.

2.1.6.3. Water use. As noted above, the principal use of water in mining and milling uranium is as makeup fluid for the wet acid-leach process employed in concentrating the uranium for shipment. The residue from the concentration plant is slurried in wet form to a tailings pile, where the water is decanted and commonly rerouted to the mill for recovery of additional uranium. Eventually, any surplus water is disposed of by evaporation. Another element of consumption, which is extremely variable, is the evaporation of water pumped from underground mines in permeable, uranium-bearing sands. This water, commonly far in excess of mill demand, is simply wasted by evaporation, so as to avoid disposing of it to streams, which would become contaminated by the radioactive components.

2.2. TRANSPORTATION

Rarely does nature accommodate man by placing natural resources at the site of ultimate use. In the case of fossil fuels, and to a lesser extent nuclear fuels, planners must decide whether to convert the raw energy materials to other forms of energy at the mine site or to transport the raw materials to more convenient sites for conversion. The decision depends on the economics of transportation of the raw materials and finished products, water availability, environmental and social impacts of the plant siting, and a number of other factors.

Because of energy losses in high-voltage transmission of electricity, mine-mouth power plants are not necessarily the most cost-effective location for generating electricity. Other factors being equal, it is generally more efficient to transport the energy fuels for electric-power plants to load centers rather than to accept the energy losses that are unavoidable in long-distance transmission lines.

Transport of energy fuels is generally by rail, water, or pipeline. In the case of oil and gas, pipelines generally have an advantage, although for bulk transport over long distances tankers are the most economical form of transport for crude oil. An example is transport of crude oil from the Gulf of Mexico to the North Atlantic Coast of the United States, which is handled mainly by sea-going tankers. In the case of uranium, oil shale, and tar sand, which occur in low concentration in nature, it is clearly most advantageous to concentrate the ore at the mine site and to ship a concentrated form to refineries nearer the market for final processing. In the case of coal, rail, water, or pipeline may be used depending on geographic considerations.

Water availability in many cases plays an important role in plant-siting decisions. Where the energy fuel is mined in an arid area, the lack of water for cooling and chemical processes may tilt the decision toward siting the refining facility or power plant near an ample supply of water and moving the energy fuel to the water. This latter consideration plays a major role in the siting of coal-fired steam-electric plants, where the weight of water consumed in cooling is six to seven times that of the coal burned. Particularly if water must be lifted to considerable heights, it is usually less expensive to ship the coal to the water supply rather than to accept the high cost of pumping water uphill.

2.2.1. Water use. The only element of the transportation sector having significant water demand is that of moving coal by slurry pipeline. In this technology, finely ground coal is mixed in about equal weight with water and then pumped through a large-diameter pipeline in a manner similar to the transport of crude oil.

Slurry lines have been used for many years throughout the world to transport matallic ores over considerable distances, and a coal-slurry line has been moving coal over a 440 km distance from a coal mine in northeastern Arizona to an electric generating station in southern Nevada in the United States since 1970. The success of the Black Mesa Pipeline, which transports 4.3×10^6 tons of coal per year to the Mohave Power Plant, has stimulated the planning of a number of other slurry lines in the United States (Figure 2.2.1).

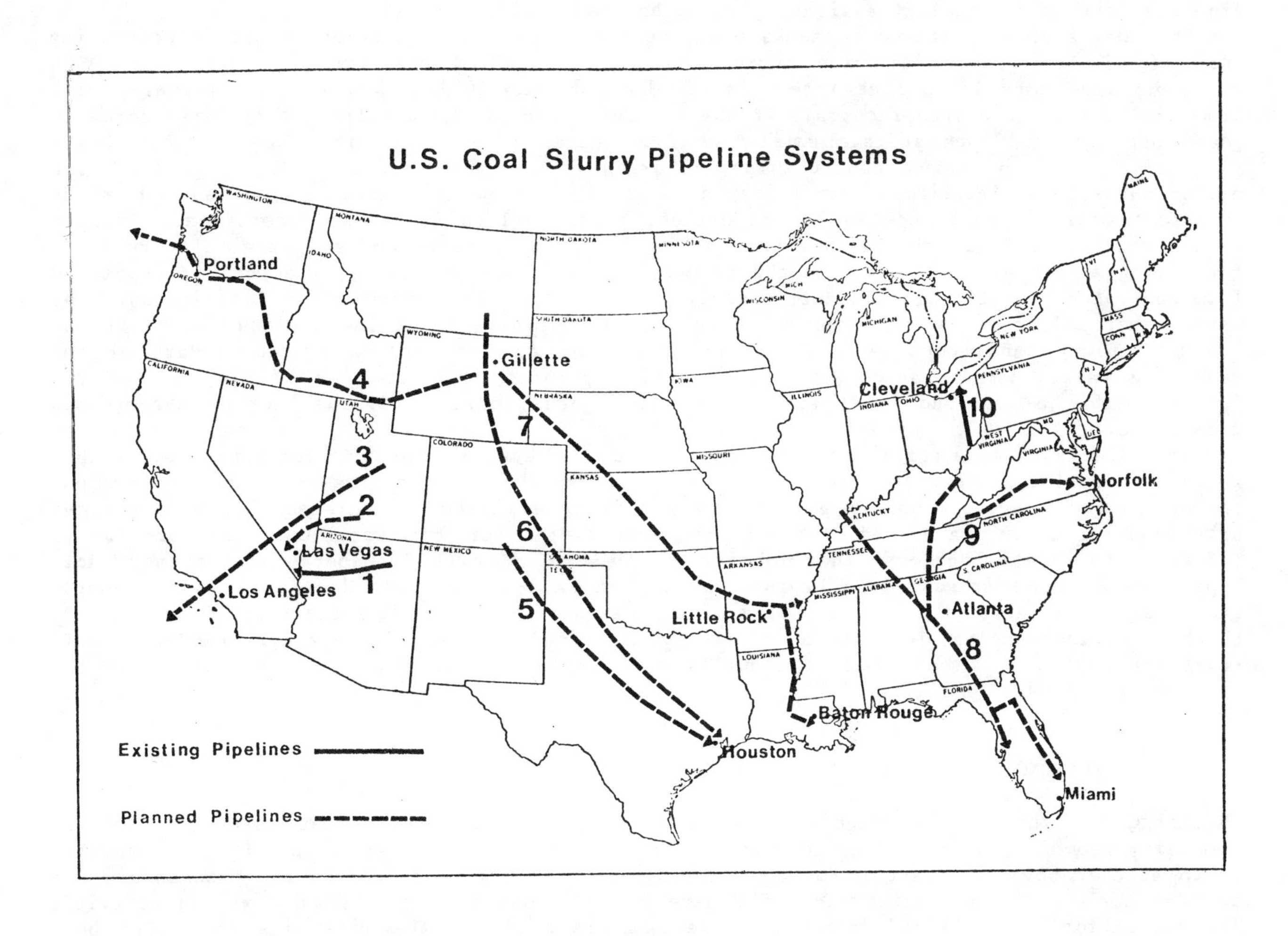

Pipeline System	Length (km)	Annual Capacity (million metric tons)
1. Black Mesa	440	4.7
2. Allen-Warner Valley	295	11.4
3. Pacific Bulk	1,048	9.8
4. Northwest Integrated Coal Energy Sys.	1,774	24.6
5. San Marco	1,452	9.8
6. Texas Eastern	2,032	21.6
7. Energy Transportation Systems, Inc. (ETSI)	2,237	24.6
8. Continental	2,419	22-54
9. Virginia Electric & Power Co.	565	4.9
10. Ohio	174	1.3

Figure 2.2.1. Existing and proposed coal-slurry pipelines in the United States, 1981. (Source: Slurry Transport Association.)

In the slurry technology, coal is assembled from a mine or group of mines at a single point, where mixing, cleaning, or other beneficiation may be done, and where the slurry is prepared. Preparation begins with impact crushing, followed by the addition of water and further grinding to a maximum particle size of 3 mm. Water is added to form a mixture that is about 50 per cent dry coal by weight. Liquid requirement is reduced to the extent of the initial water content of the coal. Water is the liquid generally used, but oil, methanol, or liquified carbon dioxide, if available, could be used in lieu of water.

The coal slurry is stored in tanks equipped with mechanical agitators to prevent settling until it is introduced into the buried pipe and propelled at about 2 m per second by reciprocating pumps located at intervals of about 80 to 240 km, depending on terrain, pipe diameter, and other design factors. At the receiving end of the pipeline, the slurry normally goes into agitated tank storage until needed. There are many possible destinations: the slurried coal can be pumped aboard ships or barges for further transport, fed into synthetic oil or natural-gas facilities, or delivered to an electric power plant. If oil is used as the transport medium, the blend can be fed directly into boilers for steam generation. In high concentration, coal slurried with water can be fed directly into boilers originally designed for firing with heavy fuel oil. In the case of the Mohave Power Plant, the coal is separated from the water by settling and centrifuging, with the fine coal removed by flocculation. The water separated from the coal represents only about one-seventh of the cooling demand of the 1,500 MW plant, and about half the water is routed to the cooling system. Part of the remaining water, which cannot be economically separated from the coal fines, enters the furnace with the coal and is exhausted with flue gases; part is disposed of by evaporation from ponds.

The water used as a carrier for the coal is not consumed in the transportation, but rather in the conversion process that follows. Nevertheless, the water represents a consumptive use to the area of origin, albeit a gain to the receiving end. For this reason, many objections have been raised in the United States to the export of water from water-deficient areas. In deference to such objections, the planners of the ERSI (Energy Transporation Systems, Inc.) Pipeline, which is scheduled to begin construction in 1983, plan to lift their water supply from the Oahe Reservoir on the main stem of the Missouri River and transport it some 400 km to the head of the pipeline in eastern Wyoming. This pipeline (see Figure 2.2.1) is designed to transport 22.5×10^6 tons of coal annually from Wyoming, 2,200 km to Baton Rouge near the mouth of the Mississippi River.

2.3. REFINING

"Refining" as the term is used in this report includes processes that change the energy commodity from its natural state to a product more suitable for consumption without converting it to another physical state. For convenience, simple on-site concentration processes, such as that employed with uranium and coal, have been discussed as part of the mining activity. The extraction of crude oil from oil shale and tar sands at the mine site have also been covered under "Mining and Milling," because in both cases the product stream is routed to conventional oil refineries for conversion to a marketable product. Thus, refining as classified herein is restricted mainly to oil and uranium treatment. The more radical processes for converting coal to synthetic natural gas, liquid fuel, chemical feedstocks, coke, and reformed coal are discussed under conversion processes.

Water is also required for natural-gas treatment and transportation. The treatment includes processes such as desulfurization (removal of the highly toxic H_2S) and removal of other gaseous contaminants that reduce the value of the product. In addition, water is used by compressors which raise the gas pressure in preparation for pipeline transportation. Because compression is an exothermic process, considerable heat is released. This heat is normally dissipated to the atmosphere by means of wet cooling towers. Little specific information is available on these widely dispersed demands other than from the government of Canada, which estimates the total water demand as of 1980 at 17×10^6 m^3 annually. This is the equivalent of 240 liters per toe; however, little information is available on what part of the water is consumed in various aspects of treatment and transmission.

2.3.1. Oil refining

The oil refining industry began in 1854 with the construction of a small plant in Pittsburgh, Pennsylvania (USA), which distilled about 800 liters per day of crude oil that was skimmed from streams and springs near Oil Creek, Pennsylvania. After the drilling of the first oil well at Titusville, Pennsylvania, in 1879, many more wells were drilled and production increased rapidly. A number of new refineries were established to handle the new production. Initially, the market was limited to production of a cheap illuminating oil similar to coal oil.

Early refinery operations were simple, because the refineries were run mainly to obtain kerosene by simple distillation. In the early refining, the lighter and more volatile gasoline and the heavier components after the kerosene was obtained were discarded or burned. Soon the residues found a market as fuel, and by 1865 lubricants were being manufactured from crude oil. The coming of the automobile in the early 1900s provided a market for the lighter components, and demand soon exceeded the quantities that could be supplied by simple distillation. The first cracking process, which yielded more of the components used in motor fuel, was introduced in 1913.

The principal processes used in modern refineries are distillation, cracking, polymerization, alkylation, and treating and finishing (Figure 2.3.1). A topping refinery separates crude oil by distillation into gasoline, kerosene, fuel oil, gas oil, and reduced crude. Atmospheric distillation is generally the first step in the refining cycle. The crude is first heated and then introduced into a fractionating tower. The lightest vapors are drawn off at the top of the tower and are condensed as gasoline; other fractions are drawn off as side streams. Water-cooled heat exchangers condense the light components and cool the side streams. The reduced crude oil from the atmospheric distillation stage generally is further processed by vacuum distillation or steam distillation to produce lubricating-oil fractions or asphalt-base stocks. A vacuum-distillation unit is operated at a reduced pressure maintained by a barometric condenser and steam jets or vacuum pumps. This process, accordingly requires both water and steam.

In the manufacture of lubricants, the lubricating-oil base stock is prepared by vacuum distillation or by propane deasphalting. The base is then dewaxed by chilling and cold filtering or pressing. Water is used in making brine solutions in the refigerating unit used in this stage, and steam is used for cleaning filter clays, pumping, and heating.

Cracking is the process of breaking down large hydrocarbon molecules into smaller, lighter molecules, Cracking not only results in a gasoline yield of 70 to 85 per cent of the input but also improves the quality of the yield. Cracking may be accomplished by thermal or catalytic processes; the thermal processes may be in the liquid phase or the vapor phase. Thermal-cracking units operate at temperatures ranging from 427°C to 650°C and at pressures ranging from 42 to 70 kg/cm^2. Catalytic-cracking units employ a catalyst to hasten the change in molecular structure and generally operate at lower temperatures and pressure than thermal crackers.

Polymerization may be considered the reverse of cracking, because two or more small molecules are combined to form a larger molecule. The process is largely used to convert by-product gases produced in the cracking process into gasoline. Polymerization may be accomplished by thermal or catalytic processes, but catalytic processess predominate. In the alkylation process, complex saturated molecules are formed by combining a saturated and an unsaturated molecule. Generally a catalytic process is used. The alkylate product is blended with gasoline to raise the octane rating. Cracking, polymerization, and alkylation all require water for cooling and other heat-transfer operations, and all use steam for regenerating catalysts, pumping, and heating.

Treating and finishing processes include caustic treating, acid treating, clay treating, oxidation sweetening, copper sweetening, and solvent extraction and a number of other processes to remove or alter the impurities in light distillates. Water is used for making up caustic and acid solutions and for product washing. Both steam and water are used to recover solvents and to clean filter clays in lubricant treatment.

2.3.1.1. Water use. The largest water use in refining, more than 90 per cent, is for cooling purposes. Generally, water is used once where it is abundant and cheap, but where water is costly or the disposal of effluent is a problem, makeup water requirements are kept to a minimum through reuse. In one form of reuse, water is used as a coolant for processes with low-temperature demands, and then the warmed water is reused for cooling in higher temperature operations. In a closed system, water absorbs heat as it flows through condensers and coolers, and the heat is removed by evaporative cooling in cooling towers, cooling ponds, and with sprayers, and the cooled water is recirculated. Most refineries use once-through cooling for certain operations and recirculated water for others.

The conventional recirculating cooling system in a refinery uses cooling towers to transfer unrecoverable heat to the atmosphere. Evaporation and windage losses occur as water passes through the tower, and a predetermined proportion of the water, termed "blowdown," is removed to prevent the build-up of mineral content of the circulating water. The proportion of the water that is used more than once or recirculated is termed "reuse". The water continually added to compensate for evaporation and blowdown is termed "makeup water."

Next to cooling, the largest use of water in a refinery is for boiler feed to produce steam. The chief uses of the steam are for stripping, steam distillation, vacuum distillation, process heating, pumping, and generating electric power. The steam that comes

Figure 2.3.1. Night view of large oil refinery of Standard Oil Co. of California at Pascagoula, Mississippi, USA. (Photo courtesy of American Petroleum Institute.)

in contact with oil in these operations and that steam's condensate generally are too contaminated for recirculation, but the condensate from boilers and pumps is reused. Smaller amounts of water, generally on the order of 1 per cent, are used for process inputs, sanitary and fire services, and other plant services.

Most descriptions of petroleum refining in the literature include little information on water use. The following data was obtained largely from reports by Otts (1963) and Evers (1975). Otts' report was based mainly on a survey of operations at 61 refineries in the United States made in 1956. Evers' report was based on a survey of 94 refineries in the United States made in 1967. The reports are in good agreement, with allowance for increased use of recirculation during the interval between the surveys. However, the Otts report presents more direct information on the consumptive use of water in the various processes.

Otts' 1956 survey found that cooling made up 91 per cent of the demand, steam production 5 per cent, and sanitary and all other uses 4 per cent. The average overall use was 40 volumes of water per volume of crude oil, and reuse was 29 volumes per volume, or 72 per cent of total water demand. Raw water intake averaged 11 volumes per volume, and effluent 9 volumes per volume, indicating that consumption accounted for about 2 volumes of water per volume of crude processed.

Evers' 1967 survey attributed 97 per cent of all water use to cooling processes, 2 per cent to boiler feed, and 1 per cent to all other uses. The average overall use was 45 volumes of water per volume of crude, and reuse was 79 per cent. Raw-water intake was about 9 volumes per volume, but no information was given on amount of effluent discharged or water consumption. However, the 21 per cent of the raw water not reused multiplied by the 9 volumes per volume of raw water intake suggests an average consumptive use of about 2 volumes of water consumed per volume of crude input, about equal to the consumptive use reported by Otts (1963). Conversion of the volumes to metric units indicates consumptive use of water of about 1,200 l/toe for oil refining. Otts (1963) points out that the largest proportion of consumption of water in refining is accounted for in the catalytic, polymerization, and alkylation units, and that these more complex processes would tend to dominate in large new refineries. Thus, the rate of unit consumptive use can be expected to climb gradually over time with increasing efforts to maximize overall plant efficiency.

2.3.2. Uranium refining

2.3.2.1. Uranium-hexafluoride production. Yellowcake, consisting of concentrates of 75 per cent uranium oxide, shipped from the mine-site concentration facilities is routed to a chemical plant, which converts the uranium oxide concentrates and other U_3O_8 streams into uranium hexafluoride (UF_6), the form required for input to gaseous-diffusion enrichment facilities. Either of two processes are used in the production of uranium hexafluoride, the dry hydrofluor process or a wet chemical-solvent extraction process. In the dry process, the concentrate undergoes a reduction-roasting in the presence of cracked ammonia ($N_2 + H_2$) to form UO_2; the UO_2 is then reacted with fluorine in the hydrofluor process to form crude UF_6, which is then purified by fractional distillation to a form suitable for input to the gaseous-diffusion separation plant. In the wet process, the concentration is digested in hot nitric acid, and the uranium is extracted as hot uranyl nitrate. The uranyl nitrate is calcined to UO_3 and then reduced to UO_2 with cracked ammonia. The UO_2 is converted to UF_6 in a hydrofluorination process and purified as in the dry process. Finally, the refined product is shipped to an enrichment plant for further processing.

2.3.2.2. Gaseous-diffusion enrichment. Normal uranium consists of 0.711 per cent of the radioactive isotope, uranium-235, the remainder of the uranium in the UF_6 input being the stable isotope, uranium-238. To bring about nuclear fission in a light-water power reactor, it is necessary to increase the proportion of uranium-235 in the blend from 0.711 per cent to 4 per cent. This is accomplished by the principle of gaseous diffusion. In this process, the solid UF_6 is heated through its liquid phase into the gas phase. The enrichment process relies on the fact that the $U^{235}F_6$ molecule is slightly lighter and smaller than the more abundant $U^{238}F_6$ molecule. The gaseous compound is pumped through a series of many porous barriers, where at each step the gas passing through is ever so slightly more enriched in uranium-235 than at the previous step. When the process reaches the enrichment level of 4 per cent uranium-235, the UF_6 is shipped to a fuel-fabrication plant, where the uranium is converted to the oxide form, shaped into fuel pellets, and inserted into fuel rods for shipment to the electic power plant (Figure 2.3.2.2.).

2.3.2.3. Water use. Considerable heat is expended in the uranium hexafluoride conversion, so there is appreciable consumption of water, as in any thermal process. The typical-sized plant produces about 5,500 metric tons per year of UF_6, which is sufficient to supply 27 nuclear power plants. The energy equivalent of the UF_6 product is about 250 x 10^9 Kcal for each 8.4

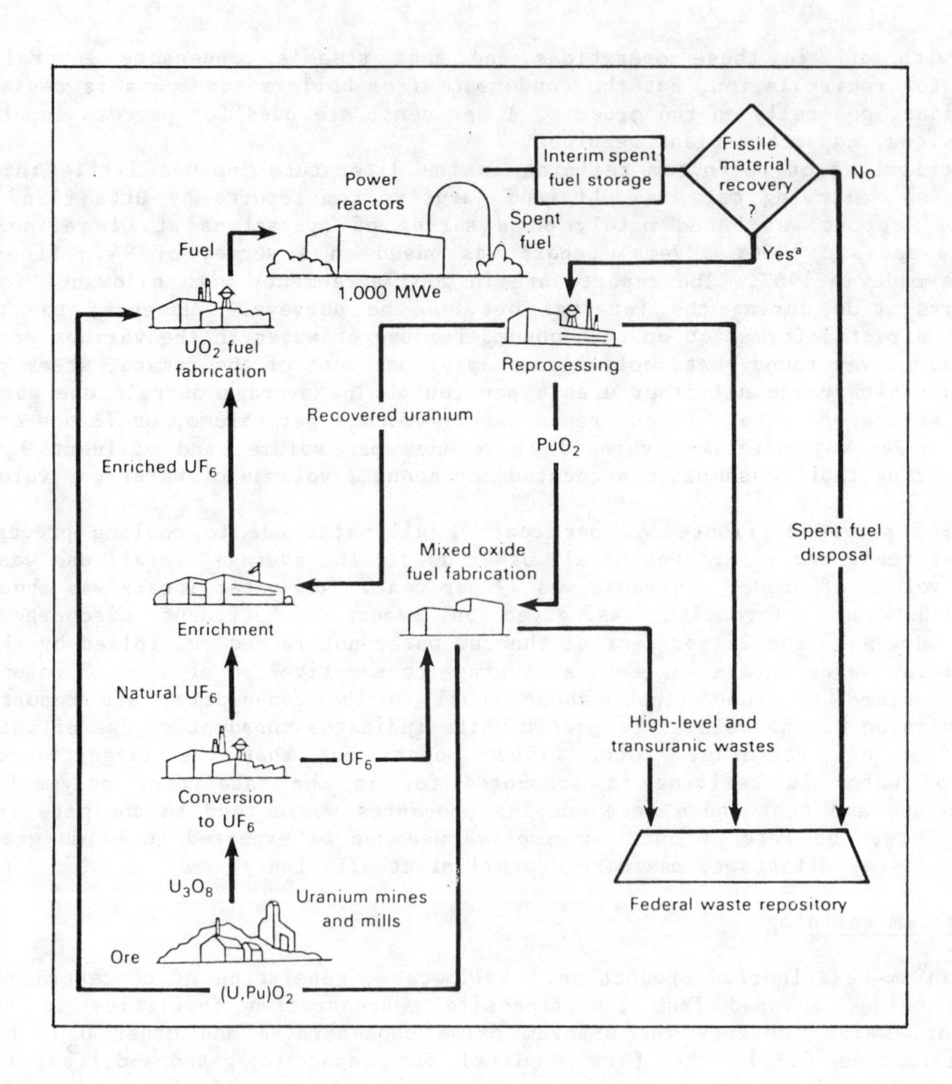

Figure 2.3.2.2. Light-water reactor fuel cycle. (Source: U.S. Department of Energy, 1981a.)

metric tons of UF_6. Thus the typical plant, rated at 117 million metric tons oil equivalent per year, consumes about 2.3 x 10^6 m^3 of water annually. The evaporative consumption of cooling water is estimated by the U.S. Department of Energy (1980) at 20 x 10^3 m^3 per Mtoe.

The gaseous-diffusion process requires a great electrical input to drive the system, and to maintain the desired operating temperatures, large volumes of cooling water are used. Indeed, next to the water consumption attributable to the uranium-milling stage, this is the largest element of water consumption in the nuclear-fuel cycle. For a modular unit of 12,000 metric tons annual capacity, the input of 9.3 metric tons of UF_6 and the output is 6.1 tons of enriched uranium in UF_6 form. The water intake is 20 x 10^6 m^3, of which 93 per cent, or 19 x 10^6 m^3, is consumed through evaporation. The unit consumptive use is estimated by the U.S. Department of Energy (1980) to be about 33 l/toe.

2.4. CONVERSION

"Energy conversion" embraces the concept of changing an energy raw material or an intermediate product into a more-usable commodity. Examples include the direct burning of coal (Figure 2.4), oil, or natural gas to convert chemical energy to electrical energy and converting the thermal energy produced in nuclear fission to electrical energy. "Conversion" can also mean changing the form of the raw material (for example, coal) into a clean, more convenient fuel (for example, gas for space heating) or even into a form of oil for further refining. Much of the current emphasis on conversion concentrates on conserving costly oil through the use of

Figure 2.4. Navajo coal-fired power plant, northwest Arizona, USA (2,200 MW) while under construction. Evaporation ponds in the foreground will dispose of 8 million m^3 per year of blowdown from forced draft cooling towers. Rail loop, left center, accommodates unit coal trains, which unload while moving. (Photo by U.S. Bureau of Reclamation.)

more plentiful raw materials, particularly coal, as a source of clean gaseous, liquid, or solid fuel that can replace oil in many applications. Commonly this involves processing the raw material near the site of production to a fuel which is then transported to the market area. Alternatively, the raw fuel may be converted to electricity near the site of production for transport in that form. A third option is to transport the raw fuel to the ultimate point of use for conversion there. Which option is chosen for a particular situation is a function of complex trade-offs among economic, water-supply, and environmental factors and a host of social factors which defy rational prediction; however, estimates of water demand are useful for planning purposes.

The processes of particular interest in the context of this report are steam-electric power generation by coal, oil, gas, nuclear, and geothermal technology; hydroelectric power generation; and conversion of coal to gas liquid fuel, or clean solid fuel. All are discussed in terms of their associated water use and consumption. For the most part, the synthetic fuel processes are not yet in operation on a commercial scale, so estimates of water use are based on planning projections rather than on experience, as in the case of electric-power production. In each mode, considerable flexibility is possible in plant design, process employed, and location of processing facilities with respect to site of extraction, source and use of water, and location of the market. It is impractical to assign rigid values of water use per unit of energy produced to all processes because of economic trade-offs, but ranges of water demand are useful for planning. However, in electric-power generation, the need for high fuel efficiency generally dictates water demand within narrow limits; accordingly, water demand for electric generation can be estimated with fair precision. The extraction of forms of crude oil from oil shale and tar sand are included by many specialists among conversion methods. In this report, however, they are treated under extraction processes, because (1) in both technologies, a form of crude oil, which must be further refined, is the output of the processing plant, (2) the processing is performed at the mine site, and (3) both technologies include a significant element of underground in-situ conversion which has more in common with mining methods than with the industrial processes described in this section.

2.4.1. Steam-electric power generation

Most of the electricity used in the world today is generated by steam driven turbines. The steam may be produced by burning coal, oil, gas, or other available fuel or by creating a nuclear fission reaction to heat water in a boiler. In other plants, steam produced from geothermal wells is introduced directly into a turbine. The type of engine employed is always one of a group termed the "Rankine cycle," in which work is done by the expansion of a heated gas against turbine blades. Fundamental to all heat-power systems working on the Rankine cycle is that a portion of the heat supplied to the cycle must be rejected. The heat rejected is primarily the heat of condensation of the working fluid (generally water, in the case of large steam generators). Fossil-fueled plants, as currently designed, waste about 65 per cent of their energy input, nuclear plants about 70 per cent, and geothermal plants 85 per cent or more. Figure 2.4.1A shows the ratio between the amount of heat rejected from a power cycle at various thermal-efficiency levels and the amount rejected from a power cycle of 35 per cent efficiency, about the average of modern steam-electric generating stations.

Practically all the waste heat dissipated from a thermal power plant is ultimately absorbed by the atmosphere, regardless of the method used for heat discharge. The methods differ in the lowest effective sink temperature that can be realized for the cycle. This sink temperature is a governing factor in the thermal efficiency of the cycle. With the exception of dry systems that operate in the fashion of an automobile radiator, all of the heat-dissipation methods result in the evaporation of water, either in a cooling tower, from sprayers, or from the surface of a pond, lake, or river, because the temperature of the heat-receiving body of water is raised (Figure 2.4.1B).

The cooling demand is governed by the thermal efficiency of the plant, which is expressed as electrical output as a percentage of the thermal energy of the fuel input to the plant. The power output of a steam turbine is greatly increased by reducing the pressure on the outlet side, by use of a condenser, which lowers the temperature of the exhaust steam, causing condensation and thus significantly reducing the pressure. The cooling capacity needed for the condensation phase of steam turbines accounts for the greatest consumption of water in the entire energy-production process. Various combinations of cooling techniques are applied to achieve maximum economy in combination with acceptable environmental effects. The cooling system is quite independent of the method of heat delivery; rather, it depends mainly on local factors, such as availability of water, terrain features, availability of land, and potential environmental impacts.

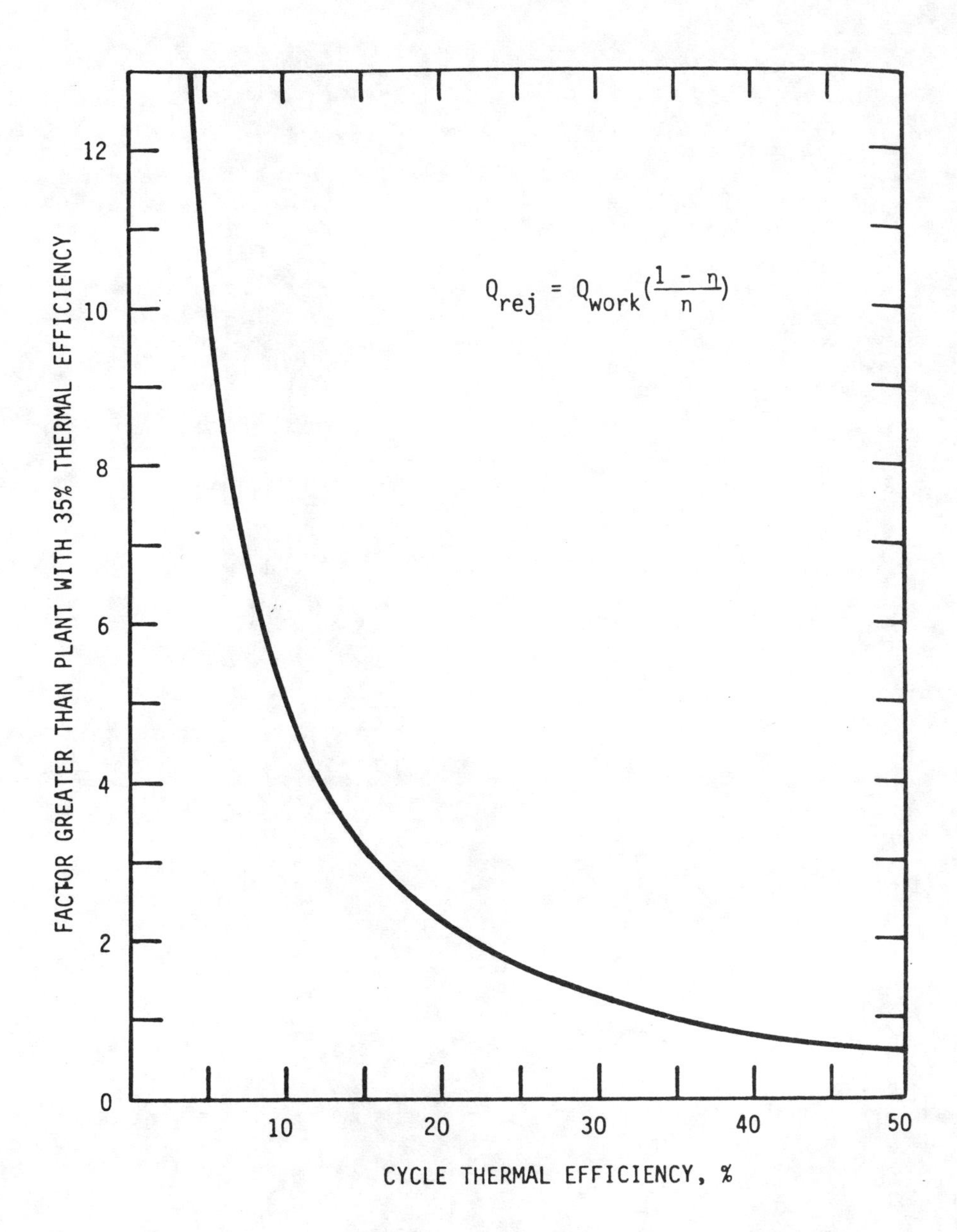

Figure 2.4.1A. Heat rejection by power-plant cooling systems vs. thermal efficiency compared with a system having 35 per cent thermal efficiency. (Source: Robertson, in Kestin, 1980.)

In modern steam-electric power plants, maximum thermal efficiency is achieved by the use of very high steam temperatures and inlet pressures. In the newer fossil-fueled plants, for example, thermal efficiency of 40 per cent is realized with inlet temperatures as high as 538°C and pressures of 246 kg/cm^2. In the cycle employed in conventional steam-electric generation, the theoretical limit of energy that can be extracted from the steam is about 40 per cent; thus, 60 per cent of the input energy at a minimum is inevitably lost unless this waste thermal energy can be put to use in some other process requiring low-grade heat.

The evaporative demand of a steam-electric generator is inversely proportional to its thermal efficiency. About 85 per cent of the energy input to the typical plant is used to drive the turbines or is disposed of as thermal waste in the form of warmed water. Present nuclear plants are less efficient than fossil-fueled plants because of safety restrictions on maximum steam temperatures. Also, nuclear plants dissipate heat almost entirely to cooling water because no flue gases are released. Accordingly, a typical nuclear plant of 31 per cent thermal efficiency releases about 50 per cent more heat to cooling water than a fossil-fueled plant of comparable capacity (Figure 2.4.1C).

Figure 2.4.1B. Four Corners power plant, northwestern New Mexico, USA. Plant is cooled by Morgan Lake, foreground, which is supplied by the San Juan River. Visible stack plumes consist mainly of steam produced from damp coal. (Photo by U.S. Bureau of Reclamation.)

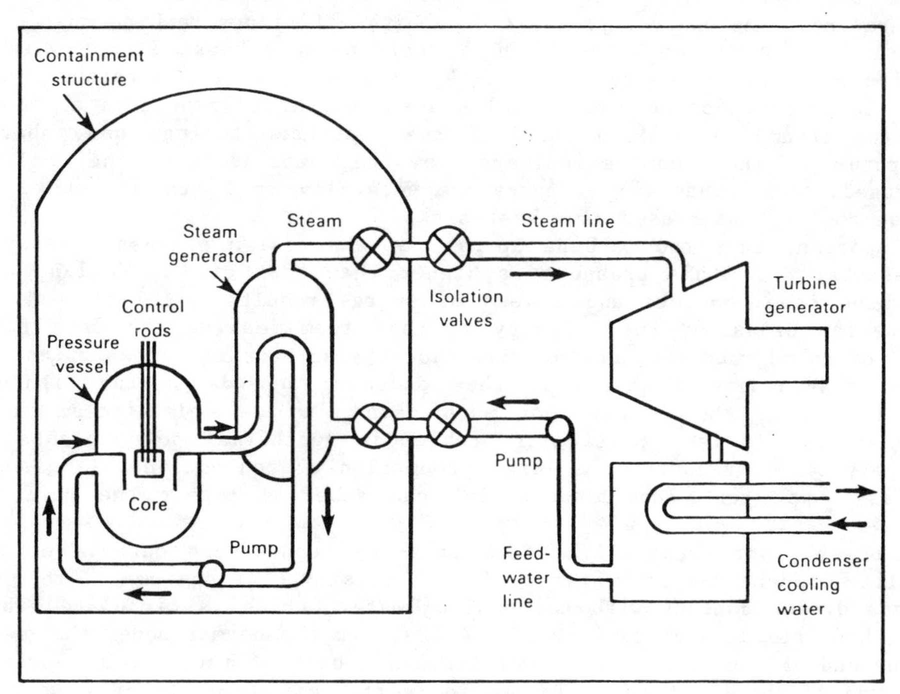

Pressurized water reactor—simplified diagram.

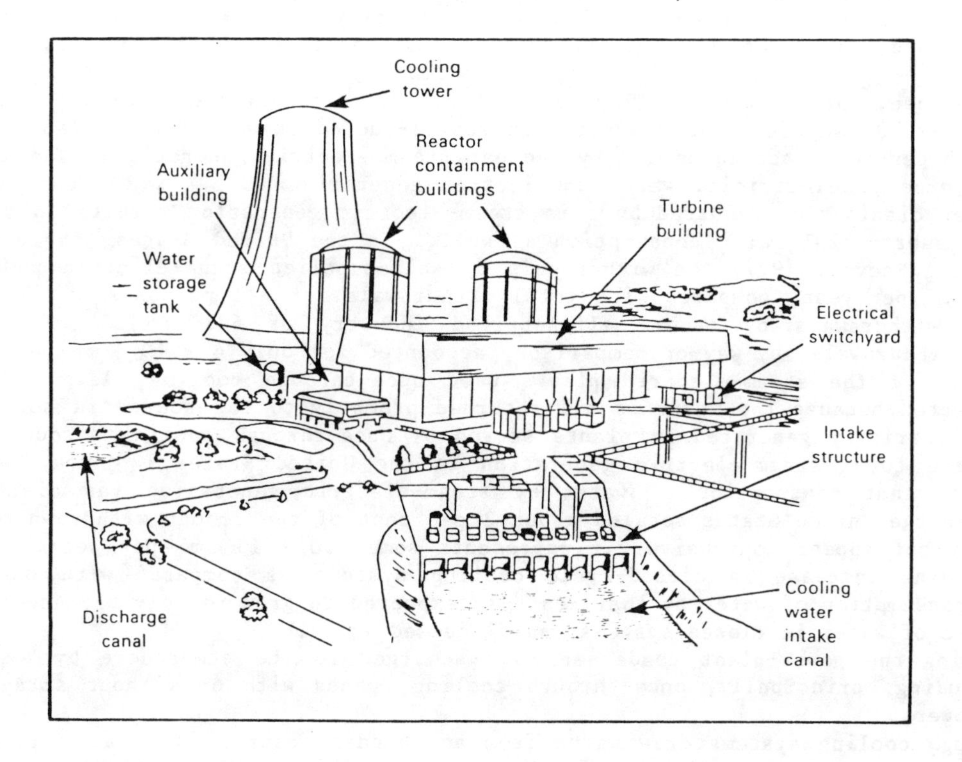

Figure 2.4.1C. Diagram of flows in light-water reactor and sketch of nuclear power-plant layout. (Source: U.S. Department of Energy, 1981a.)

With respect to consumption of water, geothermal power plants are the least efficient form of steam-electric generation. Because of inherent low temperature and pressure of the natural steam, even the most efficient geothermal plants have an overall thermal efficiency of only about 15 per cent, and many operate at less than 10 per cent efficiency.

To attain high thermal efficiency in the Rankine cycle, the turbine exhaust pressure must be as low as can be economically realized with the available cooling water or the local ambient air temperature. The exhaust pressure of a high-efficiency turbine must be well below atmospheric pressure. The vacuum is obtained by condensing exhaust steam, which in closed cycles also permits recovery of the condensate. The degree of vacuum attained depends on the turbine loading, the presence of non-condensible gases (which occur because of leakage or are introduced with the steam), the cleanliness of the condenser tubing, and, above all, the condensing temperature of the steam as influenced by the temperature of the cooling water or other heat sink used. The condensing temperature generally is 3° to $6^{o}C$ above the average temperature of the cooling water used as a heat sink.

It is also important that the turbine operate at the design exhaust pressure, commonly termed "back pressure." If this pressure is higher than that of the design, expansion of steam in the turbine is incomplete and a loss of energy results. If the back pressure is lower than the design pressure, the velocity of the steam leaving the last turbine stage exceeds the speed of sound, and the turbine will choke, also causing a loss of energy.

The amount of non-condensible gases in the condenser depends on the tightness of the system against air leaks, the presence of such gases in the inlet steam, and chemical reactions in the water. It is a particular problem in geothermal power plants because the natural steam supply commonly contains a large proportion of non-condensible gases. Even in amounts as low as 1 per cent, non-condensibles can severely impair the performance of a condenser.

Modern steam-power plants generally use surface-type condensers because of the need to recover high-quality condensate for boiler feed. In this type of equipment, the boiler steam does not come into direct contact with the cooling water, which is circulated through tubes over which the working fluid cascades (Figure 2.4.1D). In the normal mode, the coolant enters a water box on one end of the condenser, flows through a bank of tubes to the other end, then reverses and returns to the inlet end. To provide sufficient velocity of flow for effective cooling, it is necessary to use either relatively small tubing or fewer tubes of greater length. The design is a compromise between tubing that is large enough to minimize fouling and that is not excessive in length. Cooling water velocity through the tubes is usually 1.8 to 2.4 m per second; tubing of 20 to 25/ mm outside diameter is the general rule.

2.4.1.1. Water use. Of the world's 1980 total electrical generation of nearly 8.3×10^{6} GWh, 5.8×10^{6} GWh, or 70 per cent, was generated in fossil-fueled steam-electric plants, with 0.7×10^{6} GWh, or 8 per cent, accounted for by nuclear steam-electric generation. The only other significant source of electricity was hydroelectric generation at 1.7×10^{6} GWh, or 21 per cent. The heat dissipation required by this steam-electric generation results in very large water use and substantial water consumption as well. In the United States, for example, in 1975 (Murray and Reeves, 1977) the withdrawals by self-supplied steam-electric power plants was 27×10^{10} m^{3} per year, compared with total withdrawals for all purposes of 58×10^{10} m^{3} per year and withdrawals by other self-supplied industry of 6.1×10^{10} m^{3} per year. Public-supply withdrawals, by way of comparison, accounted for only 4×10^{10} m^{3} per year. Of course, as many of the steam-electric plants used once-through cooling, large withdrawals should be expected because the water is all returned promptly to the source in such systems. As of 1980, electricity generated in plants served by once-through cooling accounted for 55 per cent of the total steam-electric generation in the United States. Murray and Reeves (1977) estimate that consumption of water by steam-electric generation (as distinct from withdrawals) in the United States totaled about 2 per cent of the amount withdrawn. Although 2 per cent may not appear impressive, it represents some 3.0×10^{9} m^{3} per year. Moreover, with a continuing increase in closed-cycle cooling systems as compared with once-through cooling, the consumption of water by that means is expected to grow rapidly because of greater consumptive use of water in closed systems, as discussed later.

Heat leaving the power-plant condenser is discharged to the atmosphere by a number of methods, including, principally, once-through cooling, ponds with or without spray systems, and cooling towers.

Once-through cooling systems draw water from an abundant source (such as a river, large lake, estuary, or the sea); pump it through the turbine condenser, where the water temperature is raised 8° to $11^{o}C$, and return it to the source at a distance from the intake. This method is usually the least costly means of providing cooling for an electric plant, and until the 1960s it was the preferred method of heat dissipation in the United States. However, of the options available, once-through cooling generally is considered to have the most severe environmental impact on streams. Once-through cooling operations not only kill fish and other organisms swept into the plant with cooling water but also heat the entire body of water in

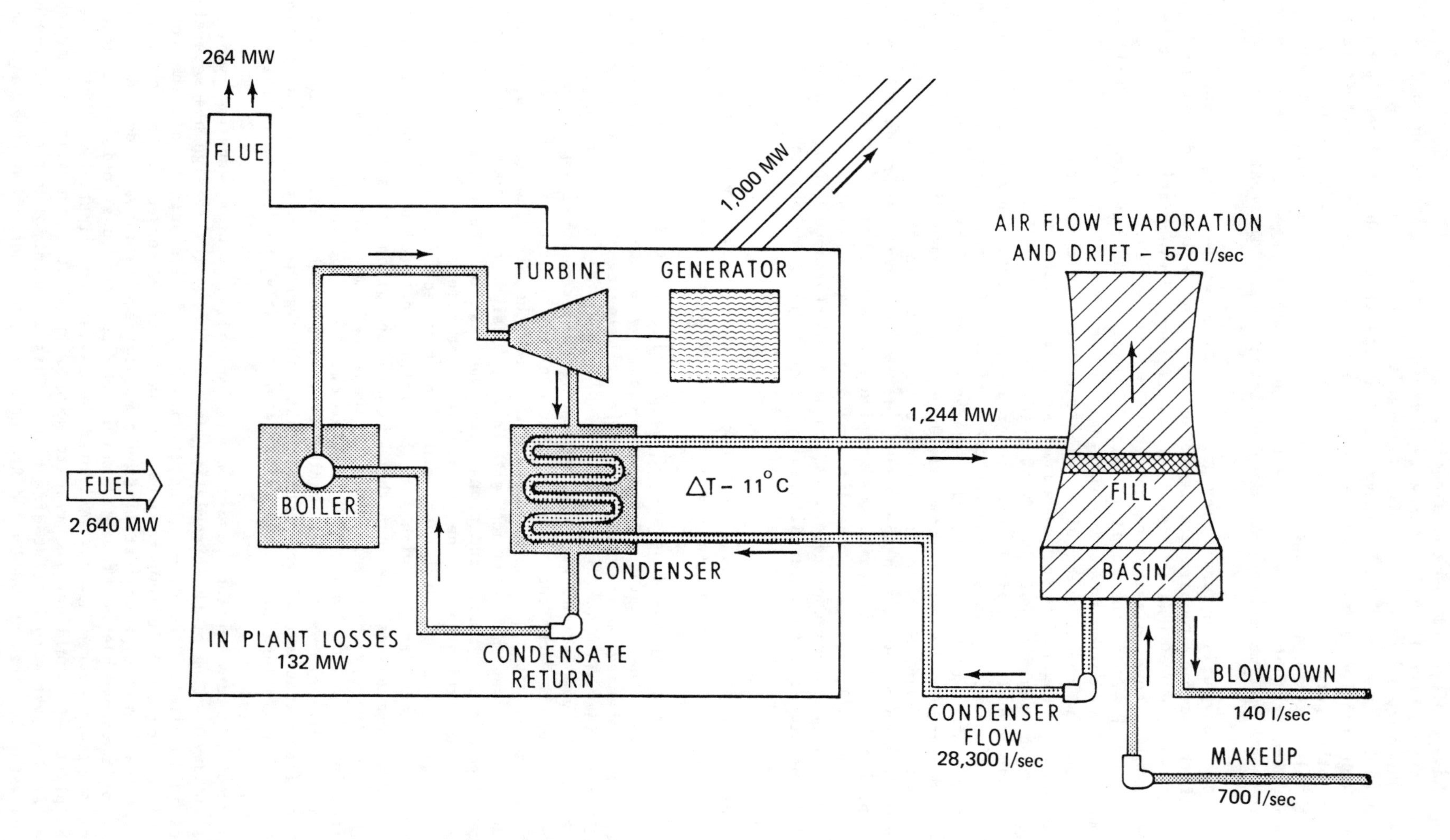

Figure 2.4.1D. Heat-balance diagram of typical 1,000 MW fossil-fueled steam-electric power plant.

the vicinity of the plant to unacceptable levels. Moreover, few locations remain where water is available all year in sufficient quantity and quality to meet the requirements of large new power plants. As a result, most plants built since the 1960s in both the United States and western Europe have employed closed-cycle cooling.

One alternative to once-through cooling is the use of cooling lakes or ponds dedicated to that purpose. However, the costs of using a relatively large land area for a cooling lake, the lack of suitable sites, and the high evaporation rate from such water bodies have combined to discourage construction of lakes solely for heat dissipation. If a storage lake is constructed for other purposes, such as flood control, navigation, water supply, or recreation then use of the lake for heat dissipation may be acceptable. One important advantage of a lake as a heat sink is that is has thermal inertia (heat storage capacity) which enables a power plant to meet daytime peak loads more efficiently and dissipate accumulated heat at night.

An open-water body will naturally evaporate 2 to 5 mm per day, depending on weather conditions. If the surface temperature is raised by using a lake to dissipate heat from a power plant, the surface evaporation increases. This induced evaporation will not be as great as that from a cooling tower dissipating an equivalent heat load, because the lake loses some heat by radiation. However, the total of natural plus induced evaporation will exceed the water consumption of a cooling tower dissipating the same heat load. From the standpoint of water conservation, the construction of a lake solely for heat dissipation is a poor choice compared with cooling towers or spray ponds. As an approximation, it is generally assumed that 4,000 m^2 of pond can dissipate 1.5 MW of heat.

Spray ponds and canals are able to achieve greater evaporation per unit of area than water bodies that rely on passive evaporation from the surface. Generally the land area required is only about 5 per cent of that of a lake of comparable cooling capacity. The cooling process, which is about 80 per cent by evaporation and 20 per cent by convection, is virtually the same as that in a cooling tower, the main difference being that with the spray system the air flow is not channeled as in a tower. In the past, spray ponds have most frequently been used in industrial applications and in small power plants. Advantages include simplicity, low maintenance, low visibility, ease of repair, and low operating cost. Disadvantages include greater land requirement than cooling towers, variable performance, limitation on approach to the wet-bulb temperature, and greater difficulty in predicting performance.

Because of environmental objections to once-through cooling systems, most large generating stations built in the United States in the past 20 years have employed cooling towers to dispose of waste heat (Figure 2.4.1.1A.). Cooling towers play a very large role in water consumption in the energy sector, so a brief description of their design and operation is given here.

The cooling tower technique consists of evaporating 1 to 3 per cent of the condenser cooling water to cool the remainder of the coolant by 8° to 16^oC. All evaporative, or wet-cooling, towers operate on the same principle of bringing the cooling water into direct contact with a moving airstream. In this mode, about 75 per cent of the cooling is accomplished through evaporation, with the remaining 25 per cent accomplished by conduction. The airstream leaves the cooling tower very close to saturation. Designs vary in the arrangement of air and water flows, the manner in which the air flow is created, and the way in which the water comes into contact with air. Design variations have advantages and disadvantages depending on local conditions, and the task of the designer is to select the most appropriate combination to fit the particular plant site.

Dry cooling is an entirely different form of heat rejection. In this mode, the coolant does not come in contact directly with air, but rather the coolant is confined within tubes over which an airstream passes. Because there is no evaporation, this mode relies entirely on conduction to raise the dry-bulb temperature of the airstream. The most familiar application of the principle is in the automobile radiator. Dry cooling is used extensively in small-scale air conditioning and in industrial processes but has found little application in power-plant cooling because of high initial cost and high power requirements; however, it offers an alternative in locations where water is very costly, because essentially no water is consumed in the process.

A variant of the dry-cooling tower is the so-called wet/dry system, which combines evaporative and dry cooling of the circulating water. The typical design has separate wet and dry sections in parallel. The dry-cooling section handles essentially all the cooling load during periods of low dry-bulb temperatures; the wet section serves as supplemental capacity during hot periods. This type of installation commonly results in water demand of only 10 per cent of that of a comparable wet tower; however, the cost is substantially greater than for either a wet or a dry system alone. As noted by Davis and Kilpatrick (1981), the cost of water at the plant site would have to rise to more than US $600 per acre-foot (US $0.50 per m^3) for the wet/dry system to be cost effective. This is nearly 100 times the present price of irrigation water in parts of the United States where this mode of cooling has been proposed as a water-conservation technique.

Figure 2.4.1.1A. Close-up of battery of forced-draft cooling towers at Navajo power plant, Arizona, USA. Large pipes on both sides of structure deliver hot water to sprinklers, which distribute water over tower fill. Air entering through louvers on sides cools falling water drops by evaporation and is sucked out by large fans at base of exhaust stacks in center of structure. Cooling water is piped from Lake Powell (Colorado River), about 2 km distant. (Photo by U.S. Bureau of Reclamation.)

Mechanical-draft cooling towers use electric fans to provide air flow through the tower fill. The units may be designed for the fans to operate in either a forced draft or an induced-draft mode, but the latter is more commonly used. Water to be cooled is pumped to the top of the tower, is distributed through headers, then cascades over a series of splash boards to the base of the tower. The direction of air flow may be counter current or cross flow to the flow of the water.

Figure 2.4.1.1B shows a typical induced-draft (crossflow and counterflow, wet mechanical-draft) cooling tower. The main advantages of counterflow towers are that the process is more efficient and can be adapted to more restricted spaces. The more widely used crossflow type has the advantages of lower air-side pressure drops and more uniform distribution of air and water streams. Each module is a separage unit with its own fan, and the louvered openings are on only two sides, permitting the modules to be arranged end to end in long rows. The sloped trapezoidal sides of the tower conform to the path of the falling water as it is pulled to the center by the cross-flowing airstream. Internal supports are made of redwood, treated fir, or iron. The splash boards are made of plastic, asbestos cement board, or polyvinyl chloride. Modern mechanical-draft towers have fiberglass fan blades up to 8.5 m in diameter and the fans are driven by 150 kW motors.

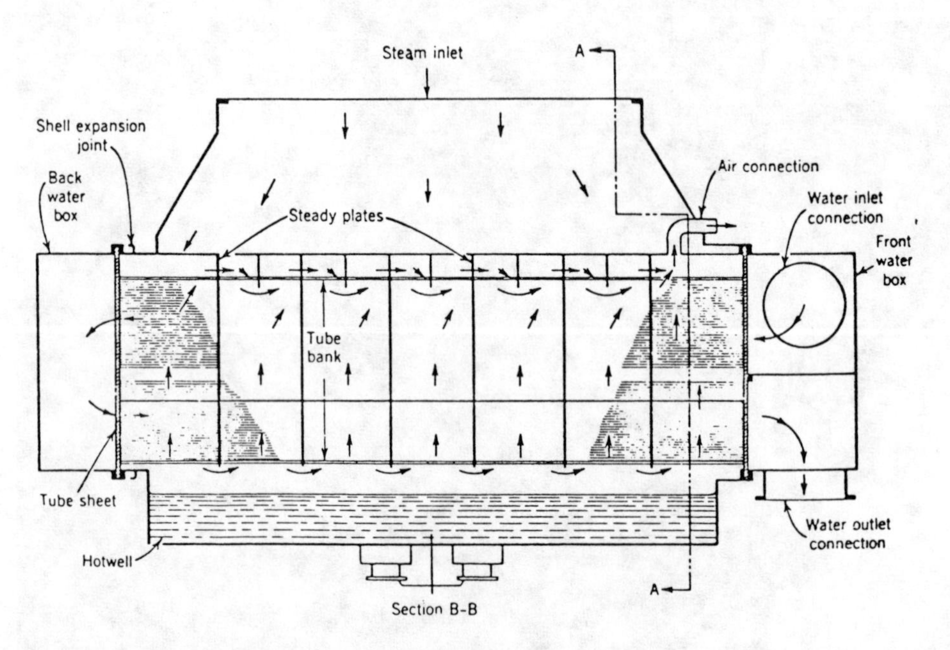

Figure 2.4.1.1B. Section through mechanical-draft cooling tower. (Source: Robertson in Kestin, 1980.)

The performance of natural-draft cooling towers is more sensitive to local weather conditions than is that of forced-draft towers. The natural-draft design was first developed in northern Europe, where wet-bulb temperatures are low and humidities are high, thus favoring that design. Moreover, peak loads tend to occur in winter, when temperatures are lowest, providing the most favorable operating conditions for the natural-draft design. In contrast, natural-draft towers are poorly suited to hot, dry areas, such as the southwestern United States, where wet-bulb temperatures tend to be high and humidity low, and peak loads occur in the summer owing to air conditioning.

The selection of natural-draft rather than forced-draft cooling towers at many sites throughout the world testifies to their advantages where circumstances are favorable. Favorable conditions include (1) low avarage wet-bulb temperature and high relative humidities, (2) a broad cooling range and a high value for approach temperature, (3) large winter loads, (4) long amortization, (5) large station size, (6) need to restrict ground-level fogging, and (7) sites where the height is not a problem. As in mechanical-draft towers, there are two basic types of natural-draft towers, crossflow and counterflow. In the crossflow, the fill over which the water cascades is located in a ring ouside the base of the tower and the inside serves mainly as a chimney. In the counterflow type, the fill is inside the base and is elevated so that the air entering the periphery moves upward through the falling water drops. This design provides more effective cooling because the coolest water contacts the coolest air. In the crossflow arrangement, air and water are distributed more uniformly, and there is less air-pressure drop across the fill. The fill consists of either splash type or film-type packing. The splash type results in bubbles continuously forming and presenting fresh surfaces to evaporation, while the film type results in a smooth film of water flowing over smooth surfaces. The hyperbolic shape of natural-draft towers matches the air flow through the units and offers strength that permits economy of construction. Reinforced concrete is the most common construction material, although other designs have proven successful in places.

An obvious difference between natural- and forced-draft towers is that, whereas the air flow in the latter is under the control of operators to a great degree, air-flow rates in natural-draft towers vary with weather and other operating conditions. Also, the water costs generally are less for the mechanical-draft design. This, however, is offset by the cost of operating fans for mechanical-draft cooling, which can total as much as 1 per cent of the plant operating costs. A variant of the natural-draft tower is the fan-assisted type. The tower fill arrangement is much the same as in a natural-draft design of the counterflow type, but fans are arranged around the periphery to force air into the tower to permit a lower height. The capacity of the tower is less subject to weather conditions and the height is less obtrusive.

2.4.2. Hydroelectric generation

In contrast to other forms of electric power, hydroelectric generation uses a renewable resource, the energy of falling water. Little water is used consumptively. The principle of water mills for the production of mechanical energy has been known for many centuries, but was first applied to electrical generation shortly before 1900. In 1980, 1.8×10^6 GWh was produced by the hydroelectric plants of the world, about 21 per cent of the total electrical generation. Hydroelectric systems form a major part of the electrical grid in several countries, notably for example, Canada, Brazil, and Norway. In many places, dams primarily for hydroelectric service serve other purposes such as flood control, low-flow regulation, and storage for municipal and irrigation water supplies.

The gross hydroelectric potential of the world is estimated by the World Bank (1980) at about 2,300 GW, of which some 1,200 GW is in the developing countries. However, as of 1980 only about 460 GW was installed, of which about 87 GW was in the developing countries. Although there are ample opportunities for increasing hydroelectric generation, particularly in the developing countries, the potential for expansion is limited by the availability of suitable sites within economic reach of power markets and by the limited availability of capital.

Hydroelectric generation can be classified into three principal technologies: (1) run-of-river plants, which use the basically unregulated flow of streams, (2) reservoir-type developments, in which water is impounded by a dam to provide a supply of water when desired, and (3) pumped-storage projects, usually operated on a daily cycle, in which water is pumped to an upper reservoir during off-peak times and the head is used to generate electricity during peak-load periods when it has greater value. In all methods, water is passed through a turbine generator combination, thus converting the mechanical energy of the water into electrical energy. In each case, the water is little affected in the process, and no significant heat is generated or dissipated; hence there is insignificant consumption of water in the generating process. If the water's surface area is increased, however, evaporation additional to the natural evaporation occurs, which represents water consumption attributable to the purpose of the change. Run-of-river developments usually result in no significnat change in surface area, and hence result in no increased water consumption. In pumped-storage projects, the increased surface area of the reservoirs is usually minimal, and thus evaporation does not represent of significant consumptive use of water. In reservoir-type developments, however, the water's surface area normally is greatly increased and this can result in significant additional evaporation over that of original land surface.

2.4.2.1. Water use. In terms of gross water use in hydroelectric-power generation, that is, the water that passes through the turbines, the figures are most impressive. In the United States, for example, the daily water use was reported to be 12×10^9 m^3 in 1975 (Murray and Reeves, 1977), about twice the average flow of the Mississipi River at St. Louis, Missouri, USA. Such astonishing figures are attributable to the fact that on highly regulated rivers such as the Tennessee, the Volga, the Parana, and the Danube the water is passed through turbines repeatedly on its way to the sea, and the repeated reuse has no significant impact on the water.

Reservoir-type hydroelectric developments, however, entail substantial consumption of water in the form of evaporation from the reservoirs used to store the water. According to calcuations by Avakyan (Davis and Velikanov, 1979), the total surface area of man-made lakes and reservoirs is more than 400,000 km^2. In the northern European part of the USSR, additional evaporation from reservoir surfaces compared with the original land surfaces is estimated to be 100 to 300 mm annually. The comparable figure for reservoirs in central Asia can be 1 m or more. In parts of the southwestern United States, potential evaporation is as much as 2 m realizable from a reservoir surface. Thus, the additional evaporation due to water storage can be very large. For example, if, as a rough estimate, additional evaporation of 500 mm were assumed for the 400,000 km^2 of reservoir surface in the world, the consumptive use would total 200 km^3. However, only a small, undetermined portion of the total additional evaporation can be reasonably charged to hydroelectric generation. First, many of the world's man-made reservoirs have no hydroelectric generation, and, of those that have, many serve other purposes in addition to electrical generation. To arrive at a meaningful estimate of consumptive use of water by hydroelectrical generation, it would be necessary to know the additional evaporation from reservoirs serving only hydroelectric generation and then to allocate the proportion of additional evaporation properly chargeable to hydroelectric generation from multi-purpose projects. This latter step would require benefit/cost analysis of individual projects, a highly subjective art, and would make the analysis a Herculean task.

2.4.3. Synthetic fuel production

Because world oil prices have remained high, nations with alternative sources of energy fuels are giving serious consideration to using those fuels as substitute sources of petroleum. The chief sources of interest are coal, oil shale, and tar sands. In each case, a liquid or gaseous synthetic fuel (commonly called "synfuel") can be produced. All conversion processes require heat and cooling, but they also require water for key chemical reactions. Although interest in synthetic fuels is now high, the processes are not new. For many decades, coal was the main source of gas for illumination and cooking in North America and Europe, until it was largely displaced in the 1930s by low-cost natural gas distributed by long-distance gas pipelines. As recently as World War II, coal served as a major source of liquid fuel for the German was effort. Indeed, the coal conversion plants now under construction and operating in the United States and South Africa are based on German designs of the World War II period.

Although oil shale and tar sand are classed by most specialists as synthetic fuels, they are discussed in this report under extraction processes for the reasons given in Section 2.3. Thus this section will be limited to the synfuel processes that use coal as the input: coal gasification, coal liquefaction, and production of clean, solid fuel.

Basically, synfuel production entails the conversion of carbonaceous material to another form. Various objectives may be met: (1) using an abundant source of carbon to produce a desired hydrocarbon product that is in short supply; (2) removing certain constituents, such as sulfur and nitrogen compounds, that give rise to air pollution when coal is burned directly; and (3) removing mineral matter that does not burn but simply forms ash, thereby producing a fuel that is cheaper and easier to handle.

The conversion of coal to synthetic fuels generally involves a process of hydrogenation, in which coal containing about 75 per cent carbon, 5 per cent hydrogen, and 20 per cent mineral matter is converted into a hydrocarbon containing a greater proportion of hydrogen, for example, methane, which consists of 75 per cent carbon and 25 per cent hydrogen. Water generally serves as the source of the hydrogen, as a coolant, and as a cleaning agent.

Published estimates of water use and consumption in synfuel processes vary widely, generally because of variation in the assumed demand for cooling water. Because there are many possible processes and many ways of combining process components in a project, this variability perhaps should not be surprising, considering that most estimates are based on paper plans, not actual experience. Furthermore, much of the variance can be ascribed to differing perceptions of the cost and availability of water supplies. Where the cost of water is high and its availability uncertain, planners have an incentive to minimize water consumption by relying on maximum reuse of water and use of air cooling to the greatest extent feasible. Conversely, where water is plentiful, particularly if the quality is poor, there is little incentive for water conservation, and both water use and consumption tend to be higher per unit of product.

Coals planned for use in synthetic-fuel production include bituminous at about 1.4 k cal/kg, subbituminous at about 0.92 k cal/kg, and lignite at about 0.75 k cal/kg. Although lignite has the lowest energy content, for some processes it is the most desirable input, because its low energy content is due to its relatively low content of fixed carbon, about 25 per cent, high content of volatile hydrocarbons, also about 25 per cent, and high content of moisture, about 50 per cent. Both the high volatile content and the high moisture content are desirable, particularly where availability of water supplies is a serious concern.

Direct or indirect hydrogenation and pyrolysis, either alone or in combination, are the main processes employed in synfuel production. Direct hydrogenation involves exposing the coal to hydrogen at high pressure. Generally, the hydrogen is produced by reacting steam with carbon char. Indirect hydrogenation is accomplished by reacting coal directly with steam or dissolving the coal in a hydrogen-donor solvent. In pyrolysis, coal is heated in an inert atmosphere and then is decomposed to yield solid carbon and gases and liquids with higher hydrogen contents than the original coal. Pyrolysis generally is a first step in most processes. Water vapor and oil fractions driven off during pyrolysis are separated out. The water, termed "process condensate," can be treated and reused.

To produce clean-burning fuels, the sulfur and nitrogen compounds in the coal must be removed. Sulfur, mainly in the form of hydrogen sulfide, and nitrogen, in the form of ammonia, are present in the gas made from the coal and in the gas produced in hydrotreating pyrolysis oils and synthetic crudes. Removal of the hydrogen sulfide and ammonia may require a liquid wash using water, and steam may be needed for heating or for gas cleaning. The demand for process water consists mainly of water for hydrogenation and for removal of contaminants.

Cooling, or the dissipation of heat, is an essential element of all synfuel processes. The heating value in the input coal that is not recovered in the synfuel product or byproducts must be transferred to the environment. Some of this waste heat is lost directly to the atmosphere, as flue gases, as water vapor from coal drying, and in other direct ways (such as convection and radiation from machinery and containers). However, the largest proportion of the waste heat is dissipated indirectly to the atmosphere through a heat-transfer surface. The heat-transfer surface may be cooled by water, which is then cooled in a cooling tower by evaporation of part of the water. Alternatively, the heat may be dissipated to a stream of air passing over a metal surface as in an automobile radiator. This latter method, which consumes no water, is termed "dry cooling," and the evaporative systems are termed "wet cooling." Under the same conditions, a dry-cooling system dissipates less heat per unit area than does a wet system; therefore, for comparable cooling capacity, a larger surface is required for dry cooling. This adds to the initial cost of the unit. In addition, a dry-cooling system cannot reach as low a temperature as a comparable wet system, resulting in lowered efficiency in hot weather. Nevertheless, in many stages of synfuel production, dry cooling is preferable to wet cooling.

The quantity of water required for wet cooling depends mainly on the overall thermal efficiency of the process employed for conversion. The overall efficiency determines the proportion of waste heat that must be dissipated. Not all the waste heat must be dissipated by cooling systems, however, because much of it escapes to the atmosphere by other means. In general, the greater the hydrogen proportion in the final product, the more processing is required, resulting in lower thermal efficiency and higher cooling requirements. Thus, coal gasification requires more cooling than coal liquefaction, which in turn requires more cooling than production of solid fuel, as shown in the table below.

Process	Carbon/hydrogen weight ratio	Process conversion efficiency (%)	Relative cooling requirement
Coal gasification	3	65-70	6
Coal liquefaction	9	70-75	3
Solid fuel production	16	75-80	1

Synthetic fuels can be produced in several ways, as shown in Figure 2.4.3. Coal generally is treated first by pyrolysis, and the resulting gaseous, liquid, and solid products are further treated. The off-gases undergo gasification or hydrogasification, and the liquids are hydrotreated to raise their hydrogen content and reduce their sulfur and nitrogen contents. If the heavier fractions are cooled instead of being hydrotreated, a clean solid fuel results. The following simplified reactions illustrate the major processes:

<u>Combustion:</u>

$$C + nO_2 \longrightarrow (2 - 2n)CO + (2n - 1)CO_2 \qquad (1)$$

<u>Gasification:</u>

$$C + H_2O(\text{steam}) \longrightarrow CO + H_2 \qquad (2)$$

Hydrogenation:

$$C + H_2 \longrightarrow CH_4 \quad (3)$$

Water-gas shift reaction:

$$CO + H_2O \rightleftharpoons H_2 + CO_2 \quad (4)$$

Methanation:

$$CO + 3H_2 \longrightarrow CH_4 + H_2O \quad (5)$$

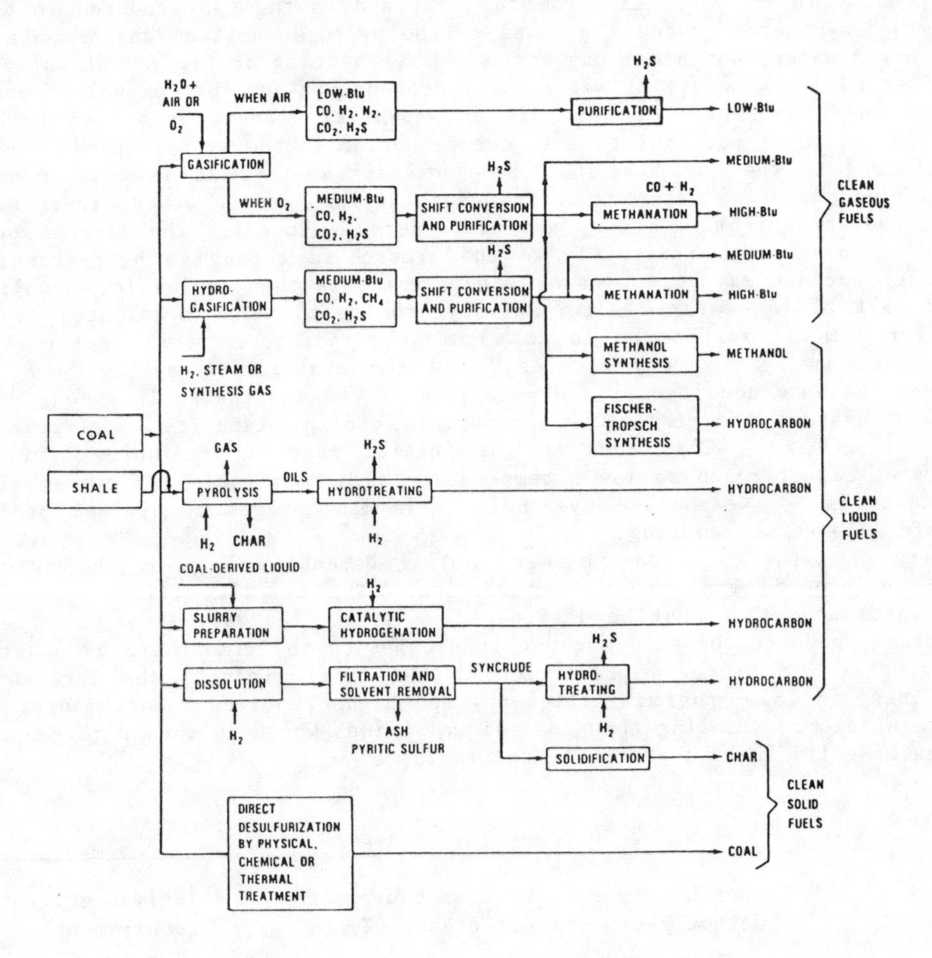

Figure 2.4.3. Conversion methods for producing clean fuels from coal and oil shale. (Source: Probstein and Gold, 1978.)

The gasification reaction is endothermic, that is, it requires heat input, usually supplied by burning coal or char (the combustion reaction, above). All the other reactions are exothermic, in that they give off heat.

The standard module plant outputs commonly used in planning synthetic fuel developments are:

Coal gasification--2.2 Mtoe (250×10^6 standard ft^3 per day) annually.
Coal liquefaction--2.5 Mtoe (50,000 barrels per day) annually.
Solid fuel--2.4 Mtoe (10,000 short tons per day) annually.

2.4.3.1. Coal gasification. Synthetic gas includes three basic products: low, medium, and high heat-value gas. The low heat-value gas, often termed "power gas," has a heating value of about 890 to 2,200k cal/m^3. It is an ideal turbine fuel, but because of its low heat value is unsuited for transport and must be consumed near the site of production. Medium heat-value gas, sometimes termed "industrial gas," ranges in heat value from 2,200 to 4,900k cal/m^3. Its main use is as a feedstock for the production of methanol and other liquid hydrocarbons, and it may be used for the production of high heat-value gas, which ranges from 8,190 to 8,900k cal/m^3. Because its heat value is comparable to that of natural gas, it is suitable for pipeline transport and can be freely substituted for natural gas.

As shown in Figure 2.4.3, synthetic gas can be produced in a number of ways. Low heat-value gas is produced by blowing the gasifier with air. This results in a high nitrogen content which has no heat value. If the gasifier is blown with oxygen instead of air, a medium heat-value gas results. If the gasifier is blown with a hydrogen-steam mixture instead of oxygen, a product higher in methane is produced.

High heat-value gas consists of about 90 per cent methane, the rest being hydrogen and carbon dioxide. A prime objective in gasification is to achieve the maximum methane level during gasification, to reduce the need for further upgrading. Of the reactions shown above, only the combustion reaction goes to completion. The composition of the gas leaving the gasifier is related to the relative contribution of each reaction to the entire process and thus is determined by the rate of individual reactions and the residence time of reactants and products in the gasifier. Reaction rates may be increased by increasing the proportion of various components in the input. Operating conditions of pressure and temperature also affect the reaction rates. For example, at low temperature, the value of n in the combustion reaction approaches 1, and thus the output is mainly CO_2. As the temperature rises, n decreases and CO is preferentially formed. Increasing the proportion of steam in the feed reduces temperatures because the gasification reaction is endothermic. If the gasifier is operated at high temperature, in the range of 1,000°C, in addition to producing a low CO_2 gas, a higher proportion of coal is converted because the fixed carbon of the char is gasified also. If the gasifier is operated at low temperature, the carbon-hydrogen reaction is enhanced, which increases the amount of methane in the output.

The off-gases from both the oxygen-blown and hydrogen-blown gasifiers are in the medium heat-value range and require further treatment if they are to be transported by pipeline. Conversion of the intermediate product takes place in two stages. First, steam is introduced into the gas in a shift converter, which uses a catalyst to speed the water-shift reaction. In this stage, the ratio of H_2 to CO is raised to 3 to 1 in preparation for the methanation stage, in which hydrogen and carbon monoxide react in the presence of a catalyst to form methane. All the reactions in the production of pipeline gas release heat. In theory, this heat could be recovered and used elsewhere in the process for pre-heating coal or producing steam; however, in practice some of the heat escapes, reducing the overall thermal efficiency.

The gases leaving a gasifier or reactor contain contaminants, such as water vapor, tar, particulate matter, and nitrogen and sulfur compounds, which must be removed before the gas can be put to use. This is accomplished in a gas purification stage employing an absorber. Between 5 and 6 per cent of the energy of the raw coal is consumed in this stage, with proportional water consumption. Most of the energy use is for boiling solvent, which is an integral feature of the absorption process, and this energy is ultimately dissipated in condensing the solvent.

2.4.3.2. Coal liquefaction. Coal can be converted to liquid end products in a number of ways, as indicated in Figure 2.4.3. The principal processes used are hydroliquefaction, solvent extraction, liquid-hydrocarbon synthesis or Fischer-Tropsch synthesis, and pyrolysis. Hydroliquefaction is the direct hydrogenation at high pressure and is the primary direct method used. In this technique, dried pulverized coal is slurried with oil, mixed with hydrogen, and fed to a catalytic reactor. The principal products are a synthetic crude or hydroliquefaction low-sulfur fuel oil and high energy-content gas. In hydroliquefaction, most of the oxygen in the feed coal is converted to water and the remainder to CO_2. The sulfur is converted to H_2S, and the nitrogen compounds to ammonia.

Solvent extraction is an indirect method of hydrogenation whereby the coal is partly dissolved in a solvent, which exchanges hydrogen with the coal. As the coal dissolves in the reactor, the ash and pyritic sulfur can be filtered out. Most of the hydrogen is exhausted in the formation of water, hydrogen sulfide, and ammonia, and the result is a low degree of hydrogenation. In some variants, additional hydrogen is added to the solvent in a separate catalytic stage, producing greater hydrogenation in the reaction with the coal.

Fischer-Tropsch, or liquid-hydrocarbon, synthesis starts with a high heat-content gas that has been through the shift reaction to produce mainly CO and hydrogen. This gas is then reacted in the presence of a catalyst to produce a variety of liquid hydrocarbons. The Sasol Plant in South Africa, the world's only commercial-scale coal-liquefaction operation, uses this technology to produce a variety of liquid fuels and petrochemicals. This process, because it starts with gas and then converts it to liquid, is inherently of low efficiency, 60 per cent or less, and accordingly is high in water consumption.

Pyrolysis, the destructive distillation of coal, yields tar, oil, water, char, and non-condensible gases. Generally, the condensates are further hydrogenated in other processes. Pyrolysis generally produces considerable water, not only from distillation of the moisture content of the coal but also because most of the oxygen in the input coal is converted to water. Water produced typically is on the order of 30 to 40 per cent by weight of the oil produced.

2.4.3.3. Solid-fuel production. In producing clean, solid fuel, the input coal is dried and then partially dissolved in a solvent that exchanges hydrogen with the coal. The solvent is then mixed with additional hydrogen and heated under pressure to dissolve the remaining carbonaceous material of the coal. The solution is then filtered and the solvent distilled off for reuse. The remaining solidified material contains less than 0.2 per cent ash and less than 1 per cent sulfur by weight; the heating value is about 1.8 kcal/kg. The product can be burned directly as fuel.

2.4.3.4. Water use. Because the overall thermal efficiency of coal-synfuel processes is relatively high, the water consumption by these processes can be expected to be moderate. Few of the processes have advanced beyond the pilot stage, so the demand for water, even on a world scale, is negligible to date. As noted earlier, generally the lower the C/H ratio, the higher the water consumption. Thus, gasification requires more water than liquefaction, and liquefaction requires more water than production of solid fuels. This is illustrated in Figure 2.4.3.4, modified after Probstein and Gold (1978), which shows typical ranges of water consumption normalized to unit heating value of the product for coal-synthetic-fuel processes. In each grouping, values are given for consumptive use by heat-dissipation systems operating at maximum, intermediate, and minimum water-cooling levels. Coal gasification ranges from 2,300 to 3,600 l/toe, liquefaction from 1,900 to 3,000 l/toe, and solid fuel production from 1,000 to 1,600 l/toe. The difference between maximum and minimum water use in each category represents a tradeoff between water costs and process costs. At the minimum level, considerable extra capital and operating cost is entailed in using air cooling to the maximum extent feasible because air cooling in most processes is inherently less efficient than water cooling. Most plants can be expected to operate in the intermediate to maximum range except where severe water shortages exist. Probably a more significant restraint on development is the high capital cost of plant construction combined with a limited market for products at the price level needed to recover investment costs.

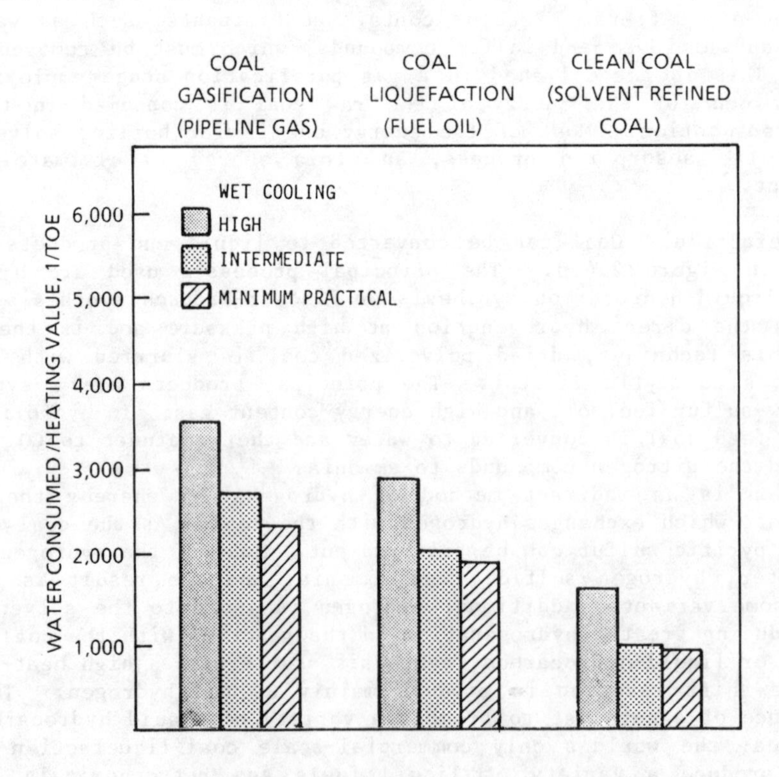

Figure 2.4.3.4. Unit consumptive use of water in coal conversion processes. (Modified from Probstein and Gold, 1978.)

As of 1983, the most serious commercial interest in the entire spectrum of synfuel processes is for production of low heat-value gas for direct input to combined-cycle electric generating plants in areas that cannot tolerate air pollution from a standard coal-fired plant. In this mode, the output of a coal-gasification plant is fed directly into a gas turbine to generate power and the hot exhaust gases produced are used to raise steam to generate electricity in a conventional steam-electric cycle. By combining cycles in this way, it is possible to circumvent the upper limit of about 40 per cent thermal efficiency of the conventional steam cycle and to achieve an overall thermal efficiency of 50 per cent or better through the use of heat that otherwise would be wasted. A plant employing the Texaco oxygen-blown gasification process is under construction for the Southern California Edison Company near Barstow, California, USA. This plant is now in operation (1983) with a 1,000-ton-per-day coal gasifier feeding gas to an existing 65 MW boiler. Late in 1985 an integrated combined-cycle generator is scheduled to begin operation at 90 MW output.

The Sasol Plants I and II now operating in South Africa are the world's largest coal-conversion facilities, with a combined input of about 15 x 10^6 tons per year of coal. When the Sasol Plant III comes on line with an additional 14 x 10^6 annual coal input in 1985, the Republic of South Africa expects to supply 50 per cent of its liquid-fuel needs from this complex.

3 Present water use and consumption in the energy sector

3.1. METHODOLOGY AND ASSUMPTIONS

To arrive at world estimates of consumption of water in the energy sector, it was necessary to assign average values of unit consumptive use to estimates of energy production of the various fuels and processes that make up the energy sector. The data on energy processes are largely taken from reports of the United Nations (1981), the Organization for Economic Co-operation and Development (OECD, 1982a), and the World Bank (1980). In keeping with United Nations and OECD practice, energy outputs and consumption are expressed in millions of metric tons oil equivalent (Mtoe) and for electrical output in gigawatt hours (GWh). Consumptive use estimates are expressed in liters per ton oil equivalent (l/toe) to avoid unwieldy quantities, and total water consumption is expressed in cubic meters (m^3) per year. Water consumptive use associated with electrical production is expressed in liters per kilowatt hours (l/kWh). A table of conversion factors is included as Appendix B to permit ready conversion to other units.

Unit-consumptive-use estimates were derived from a number of sources, chiefly reports issued by the United States and Canadian governments. Where sources differed significantly on unit consumptive use, a value for extrapolation was adopted on the basis of the rationale presented in the following section. Unit-consumptive-use estimates were then applied to energy-output data to arrive at world consumption in various elements of the energy sector. No special effort was made to segregate saline-water use from fresh-water use for two reasons. First, the only detailed reporting available was from the United States, and second, the data are flawed because the basis for reporting is salinity. Estuarine waters, which are the supply for many of the largest refineries and electric power plants in the United States, are in an ambiguous position; some such supplies are reported as saline and some as fresh. With respect to the consumption of these industries, as competition with other uses for fresh water supplies, it makes little difference whether the water is saline or fresh, because by the time the water reaches the estuary it generally is of little value for most other uses.

Because most of the world's large oil refineries are located at marine terminals on the seacoast or on estuaries, their water consumption does not compete for fresh water supplies. The U.S. Department of Energy estimates (1981a) that of the world's total petroleum production, a little more than half was transported by marine tankers in international trade. It would be reasonable to assume that all this oil is refined in plants at maritime sites. Only in the United States, Canada, and the USSR is a significant proportion of the world refining capacity situated far from the sea to serve inland markets from inland crude-oil supplies. Even in the United States, which has many inland refineries, some one-third of the capacity is supplied with saline water at coastal locations.

Steam-electric plants are generally sited near load centers to minimize transmission losses. Because seacoast cities represent major load centers in most countries, a significant proportion of steam-electric plants depend on marine or estuarine waters for supplies. As noted earlier, reporting of water supplies in this process suffers from the flaw that a sharp distinction is made between fresh and saline waters, but mixed estuarine waters are not distinguished. In maritime nations, such as Japan, the United Kingdom, and the Netherlands, the bulk of the water used in refineries and power plants is marine or estuarine, while continental nations use larger proportions of fresh water. In land-locked countries, such as Austria and Switzerland, consumptive use of water necessarily is entirely fresh water.

3.1.1. Coal mining

The various sources of information on coal-mining water uses were not in good accord. As reported by Davis and Velikanov (1979), United States estimates of unit consumptive use were 42 l/toe for underground mining and 11 l/toe for surface mining (with an additional 22 l/toe for waste disposal in each category) or 64 l/toe for underground mining and 33 l/toe for surface mining. They also indicate that water use in the USSR for underground coal mining was 293 l/toe but that this represented gross withdrawal, not consumption. Figures of 80 l/toe were cited for Hungary, where recycling was used.

Recent data published by the U.S. Department of Energy (1980) indicate unit consumptive use of 82 l/toe for surface mining and 105 l/toe for underground mining, with an additional 174 l/toe for coal beneficiation, to which 35 per cent of United States coal production is subjected. The weighted average of these estimates for the United States is 153 l/toe (exclusive of irrigation required on reclaimed lands in arid areas). It was not clear whether the USSR or Hungarian estimates included water for beneficiation, but they probably did not. Thus, it would appear that consumptive use of water in underground mining was in the range of 65-105 l/toe, not including beneficiation, and about 20 l/toe less for surface mining. Because little information was available on the practice of coal beneficiation by heavy-fluids separation except in the United States, an overall average unit-consumptive-use value of 100 l/toe was adopted for estimating world consumption of water in coal mining. Considering the uncertainty of the underlying data, precision greater than one significant figure probably is not warranted in any event.

3.1.2. Oil production

The predominant consumptive use of water in oil production is for water flooding, or secondary recovery. Water consumed in drilling is modest in most countries, even in the United States, which leads the world in depth of hole drilled per new oil discovery; only about 46 x 10^6 m^3 per year of water consumption is attributed to oil and gas drilling. Because this is a single-time use at any given site, the water consumption has little impact on local or regional water balances.

The water consumed in water flooding, in which water is used to displace oil from the producing formation, is of major significance. Estimates cited by Davis and Velikanov (1979) indicated a requirement of 1-3 m^3 water per metric ton of oil produced from secondary recovery in the United States. USSR data indicated 2.5-3.5 m^3 per ton produced for all water uses, including drilling, flooding, gathering, and treatment of oil. This latter figure, however, appears extremely high for consumption associated wtih production only, and treatment may include processes that are grouped in refining in this report. In principle, water required for flooding should, under equilibrium conditions, equal the volume of oil displaced, although in the initial phase of a water flood, additional water could be required to saturate parts of the producing zone that had been drained by production.

In this study, heavy reliance was placed on data supplied by the Inland Waters Directorate, Environment Canada, indicating annual water use for oil and gas production (exclusive of gas treatment) of 34 x 10^6 m^3 as of 1979, which was associated with 1979 oil production of 85 Mtoe. Information received from the American Petroleum Institute (John Oshinski, 1981, oral communication), based on a sampling of major oil producing states, suggests that about 35 per cent of 1980 United States oil production was from secondary-recovery projects. Because Canadian production is managed in much the same fashion and by the same producers as United States production, it is reasonable to assume about the same production ratio, or about 30 Mtoe per year for Canadian secondary production. This suggests a unit consumptive use of water of 1,100 l/toe, in good agreement with the assumption of volume for volume displacement of oil by water. This figure has, therefore, been used as the consumptive-use value for oil production in Table 3.3A.

In addition to consumption for secondary recovery, in the United States, the water consumption for tertiary production, chiefly for steam injection, must be added to arrive at total water consumption in oil production. Some 15 Mtoe, representing 3.6 per cent of United States oil production, was from steam-injection projects in 1980 (Crouse, 1981). The U.S. Department of Energy (1980) estimates the unit consumptive use of water by this methodology to be about 4,900 l/toe for generating process steam, most of which is condensed in the formation as it displaces oil. Because steam-injection methods are the principal source of oil in tertiary production, 4,900 l/toe has been used in Appendix C for the unit consumptive use by this type of production. Some 3.7 Mtoe annual production was estimated from miscible-flooding techniques in the United States in 1980 (Crouse, 1981). This also is a form of tertiary recovery; however, because the amount makes up less than 1 per cent of total United States production and the water consumption should be about the same as in secondary recovery, water consumption by this method has been arbitrarily included in that for secondary recovery.

In arriving at water consumption due to oil production in the United States and Canada, the amounts shown in Table 3.2 were calculated directly. Thus, United States water consumption for oil production as of 1980 was computed as the sum of that required in secondary recovery (186 x 10^6 m^3), in tertiary recovery by steam injection (74 x 10^6 m^3), and in drilling (46 x 10^6 m^3), or about 300 x 10^6 m^3. Water consumption due to oil production in Canada was taken to be the 34 x 10^6 m^3 reported by the Inland Waters Directorate, Environment Canada. This, of course, ignores tertiary production and water required for drilling in Canada, but it is believed that the amounts involved are within the rounding of numbers, because both amounts are relatively lower than in the United States.

Detailed information permitting water-consumption estimates for other countries was not available. Accordingly, for the rest of the world, it was conservatively estimated that 10 per cent of the production was from water flooding, and the quantity of water calculated in this manner covered other uses in oil production as well. Totals shown for the world and OECD oil production reflect the fact that the United States and Canada were calculated differently.

3.1.3. Natural gas processing and transmission

As noted in Section 2.1.2, natural gas production per se has minimal water requirements; however, as observed in Section 2.3, "Refining," much natural gas requires treatment before it can be distributed, and there are substantial water uses involved in its transmission, for example, for compressor-station cooling. THe 46 x 10^6 m^3 of water per year included in consumptive use for drilling included in the United States total under the estimates for oil production includes the requirement for gas drilling also, so in the gas sector, only consumption in processing and transmission remain unaccounted for. The only specific estimate available on this sector is that of the Inland Waters Directorate, Environment Canada, of 17 x 10^6 m^3 for gas processing in Canada as of 1980. This compares with gas production of about 70 Mtoe in Canada in 1979, suggesting consumptive use of about 240 l/toe. Considering that natural-gas treatment commonly requires removal of water vapor, hydrogen sulfide, ammonia, carbon dioxide, and other non-condensible gases, sophisticated chemical processes, such as those used in certain steps of coal gasification are involved (Figure 3.1.3). As noted in Section 2.4.3, the gas purification step in coal gasification is energy intensive, so it is not unreasonable to assume substantial cooling requirements and consumptive use of water in natural-gas processing also.

Compression of gas for high-pressure transmission is an exothermic process, and the heat emitted must be dissipated to the atmosphere, generally by wet cooling. Accordingly, substantial water consumption can be assumed. However, lacking further specific information, for the purpose of this study, the Canadian estimate has been accepted as applicable elsewhere.

3.1.4. Oil refining

Oil refining is a highly complex chemical-engineering operation. Accordingly, it requires a very large investment in facilities, and economy of scale becomes most important. As a result, the refining industry in most countries is concentrated in a few centers, and individual refineries have a capacity of 12 million tons or more of crude input per year. Because of the petroleum industry's heavy reliance on marine tankers for transport, most refining is done at a few world centers at seaports or estuarine sites. Accordingly, much of the water demand is met by marine water, although as noted earlier in this chapter estuarine waters make classification difficult.

Water is consumed for a great variety of cooling purposes and for process input as well. Where water is plentiful, once-through cooling commonly is used and the intakes are large. However, the processes employed in oil refining have high overall thermal efficiency, in the neighborhood of 70 per cent, so where water is scarce or expensive, water intake can be greatly reduced and much of the cooling demand can be met by air cooling. However, because the greatest proportion of the world's refining is done at marine terminals, water supply rarely is a critical problem.

The high capital investment in refineries demands the most efficient design; consequently, most large modern plants operate in the same thermal efficiency range, and their consumptive use of water, regardless of whether the plants maximize reuse of water or not, is fairly uniform. Accordingly, the unit-consumptive-use estimates shown in Table 3.2A of about 1,200 l/toe are considered appropriate for extrapolation world-wide and have been used in Table 3.2B.

Figure 3.1.3. Typical gas-processing plant. Sulphur recovery unit, left foreground, produces about 150 tons per day. Plant also separates out propane and butane in processing 3 million m^3 of natural gas per day. (Photo by U.S. Bureau of Mines.)

3.1.5. Nuclear fuel cycle

The nuclear fuel cycle as defined by the U.S. Department of Energy (1981a) consists of uranium mining, ore-milling including concentration, conversion of concentrate to uranium hexafluoride, uranium enrichment by the gaseous-diffusion method or alternatively by the gas-centrifuge method, fuel fabrication, and fuel reprocessing where appropriate. As practiced in the United States, the largest unit-consumptive use in these steps is for gaseous-diffusion enrichment, at 490 l/toe. In the gaseous diffusion process, uranium hexafluoride is gasified, requiring high thermal input, and the gas is forced through a series of filters, also requiring large energy input. The cooling requirements, normally met by wet cooling, accordingly, are high. If gas centrifuging is substituted for gaseous-diffusion enrichment, only 73 l/toe is required for this step, a reduction of 85 per cent in unit-consumptive use of water.

Next to enrichment, the most water-consumptive phase of the cycle is the uranium-milling step. This high water consumption is due to the use of wet milling and concentrating processes, which yield large volumes of fluid waste. This waste is generally disposed of on-site by evaporation to avoid contamination of surface and ground waters with radioactive effluents. The unit-consumptive use of this step is estimated to be 410 l/toe.

For all steps, from mining through delivery of fuel to a nuclear power plant (Figure 3.1.5), the unit-consumptive-use values aggregate to 960 l/toe under the gaseous-diffusion enrichment option and 540 l/toe under the gas-centrifuge option. Most of the world's current production for nuclear power plants uses the gaseous-diffusion method. Power-plant-fuel reprocessing, which is not practiced in the United States currently, requires an additional 20 l/toe.

Because most of the nuclear power plants in operation in the world at this time are supplied with fuel by United States suppliers (or use comparable methods for the manufacture of fuel), the aggregate value above has been adopted for computing water use thoughout the world; however, it has been raised to 970 l/toe to make allowance for fuel reprocessing where carried on.

Canadian nuclear power plants (CANDU reactor design) use unenriched uranium fuel, and thus no water is consumed in the enrichment process. This water saving, however, is more than offset by a 1980 requirement of 4 million m^3 per year reported to Unesco by Environment Canada for production of heavy water, which serves as the primary coolant and moderator in CANDU reactors. Accordingly, the water consumption shown in Table 3.2B for Canada under "Indigenous production, Nuclear," was calculated by adding 1.2 million m^3 for mining and milling consumption to the 4 million m^3 for heavy-water production.

3.1.6. Steam-electric power generation

Because steam-electric power generation makes up the largest water-consuming element in the entire energy sector, it merits particular attention. Steam power can be generated from a variety of sources--coal, oil, gas, nuclear, and geothermal, representing nearly all of the world's present generating capacity. However, any fuel can fire a steam boiler, if available at a competitive price. Thus, wood has long been used to fuel boilers and, in favorable locations, may gain in usage for electric-power generation. Also, in sugar-growing areas of the world, bagasse, the waste left after sugar syrup has been squeezed from cane, has long been used to raise steam for boilers to operate sugar mills and rum distilleries.

Most boiler systems used for electric-power generation are of similar standardized design and operate in much the same fashion. With respect to water use and consumption in the steam cycle, it makes little difference whether the fuel is coal, oil, or gas. Light-water nuclear plants, however, are subject to special safety criteria for maximum water temperatures and thus operate at lower efficiency than comparable fossil-fueled systems. Geothermal plants are limited by the low temperature and pressure of natural fluids and require specially designed low-pressure turbines that operate at substantially lower thermal efficiency than higher pressure systems. These different levels of thermal efficiency have a major impact on water consumption, because unit-consumptive use is directly proportional to overall thermal efficiency.

As an approximation, water consumption in power-plant cooling can be calculated using graphs such as Figure 2.4.1A. However, for more precise results it is important to have information on local weather parameters, especially the average wet-bulb temperature at the plant site. Figure 3.1.6, modified from Probstein and Gold (1978), shows the relationship between wet-bulb temperature and evaporation in the principal types of power-plant cooling systems. The lower water consumption achieved through the use of once-through cooling is quite evident. Designers of power-plant cooling systems, of course, tailor the design to the conditions at the specific site. However, the problem for this report is quite the opposite, that is, to select average consumptive-use values that will approximate world averages and can be extrapolated broadly.

Figure 3.1.5. Inland nuclear power plant (918 MW) with two natural-draft cooling towers dwarfing reactor housing, cylindrical structure in foreground. Arroyo Seco plant, near Sacramento, California, USA. Cooling-water supply is delivered by Folsom South Canal of U.S. Bureau of Reclamation from Folsom Reservoir on the American River. (Photo courtesy of American Petroleum Institute.)

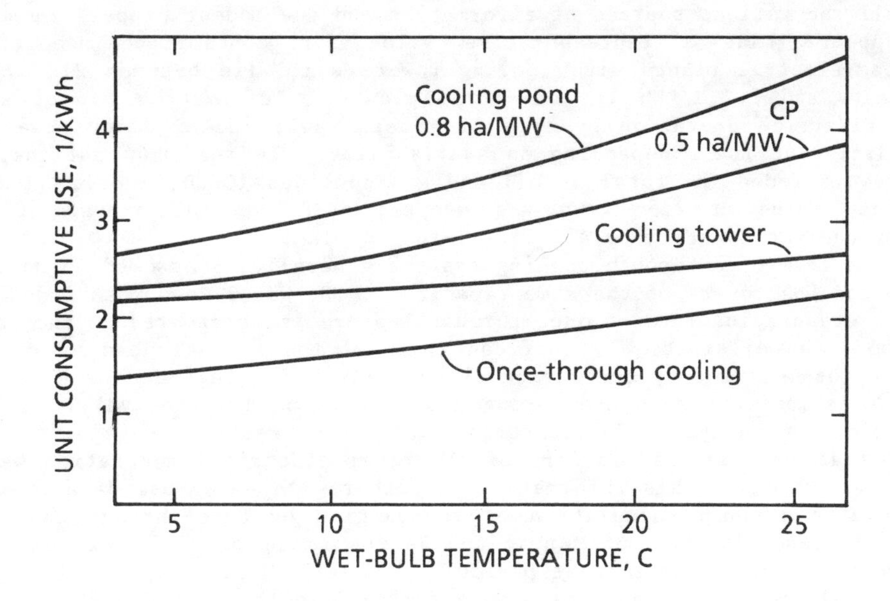

Figure 3.1.6. Relationship between wet-bulb temperature and unit consumptive use by evaporation, by type of cooling system. (Modified from Probstein and Gold, 1978.)

A number of sources were consulted for information. One of the most extensive studies of evaporation from steam-electric power-plant cooling systems was that of Hu, Pavlenco, and Englesson (1978) done on behalf of the U.S. Environmental Protection Agency. Average evaporation from cooling towers, cooling ponds, and once-through systems was computed for 37 cities in the United States, under different evaporation models and design assumptions. Consumptive-use values for cooling-tower systems ranged from 1.2 l/kWh at Bangor, Maine, a cool maritime location, to 3.0 l/kWh at Albuquerque, New Mexico, a warm arid location. The lowest value was for a plant with cooling towers, without a makeup pond and with the blowdown returned to the source, applying the Brady evaporation model. The highest value was for a plant with cooling towers, with a makeup pond and with blowdown retained, applying the Harbeck evaporation model. A value that seemed more typical of the major load centers of the United States and northern Europe was that of a hypothetical power plant at Columbus, Ohio, USA, using cooling towers, with a mixing pond and with blowdown retained. The calculated evaporation using the Harbeck evaporation model under these conditions was 2.3 l/toe.

The U.S. Department of Energy Environmental Information Handbook (1980) presents a method for calculating water losses from typical power-plant cooling towers in the eastern United States that is based on the assumptions that 50 per cent of the heat released from coal combustion is dissipated to the circulating water system and that the water heat of vaporization is 1,050 Btu/pound of water evaporated. Under these assumptions, the calculated evaporative loss is 2.4 l/kWh. However, if a value of 1,390 Btu/pound is used for the water heat of evaporation for the eastern United States, as recommended by Probstein and Gold (1978, p. 51), and a more precise value of 54 per cent (assuming 36 percent thermal efficiency and 85 per cent of waste heat dissipated to the cooling system) is used for energy discharged by the cooling system, the recalculated evaporation loss is 2.2 l/kWh.

Examination of Figure 3.1.6 suggests that at average annual wet-bulb temperatures in the range of 5°-10°C, which would embrace most of the industrial areas of North America and northern Europe, the evaporation losses from cooling towers would average about 2.3 l/kWh. Evaporation from cooling ponds and once-through systems is more sensitive to changes in the wet-bulb temperature in the 5°-10°C range, but if 7.5°C is selected as a representative value, it can be seen that the loss from once-through cooling would be about 1.5 l/kWh, and from a large cooling pond would be about 2.8 l/kWh, 65 per cent and 120 per cent of cooling-tower loss, respectively.

Coal-fired power plants generally have other consumptive uses of water in addition to cooling--for ash quenching and disposal and for sulfur-dioxide control where used--which add about 0.5 l/kWh to the evaporative requirement. In addition, all plants with closed cooling systems have a need for cooling-system blowdown, generally assumed to be 10 per cent of the evaporation requirement; however, because this is discharged to surface streams in temperate climates, it is not counted as consumptive use.

Considering all the various sources of information and the modest dispersion among various estimates, it appears that a representative value for consumptive use of water by fossil-fueled steam-electric plants with cooling towers would lie between 2.2 l/kWh in oil- and gas-fired plants and 2.7 l/kWh in coal-fired plants. Information is not available to break down world electric generation by type of fossil fuel; indeed, many power plants are designed to use different fuels depending on availability. In the OECD nations, coal-fired generation slightly exceeded the total of oil and gas combined (OECD, 1981). Accordingly, a unit-consumptive-use value of 2.5 l/kWh was adopted as a general average for computing consumptive use by cooling towers from all fossil-fueled plants.

Evaporative loss from once-through cooling systems generally is assumed to be about 50 per cent of that from cooling towers of the same capacity, (U.S. Water Resources Council, 1968, p. 4-3-2). The lower evaporation rate of once-through systems is attributed to the fact that the receiving water body can dissipate a significant part of the thermal load by convection and radiation to the atmosphere and by conduction through its bed. This can have a major impact on overall water consumption where once-through cooling is used extensively. In the United States, for example, an analysis of Department of Energy records on electric-power-plant cooling indicated that in 1980, 55 per cent of the total electrical generation was served by once-through cooling systems. This information, together with water-use data from Murray and Reeves (1977), permits a rough calculation of consumption by once-through and closed-cycle cooling systems. Because the data of Murray and Reeves is as of 1975, it was necessary to adjust the more recent information to that period.

Using OECD thermal-electrical generation data for 1975 in the United States of 1.877×10^6 GWh, it was assumed that 60 per cent of the generation was served by once-through cooling systems. This percentage was assumed because most new additions to generating capacity (6.3 per cent growth rate, 1975-1980) employed closed-cycle cooling. Thus, generation by closed-cycle plants was calculated to be 0.751×10^6 GWh, and by once-through cooling plants, 1.126×10^6 GWh. Water consumption by thermal-electric plants was reported by Murray and Reeves to be 3×10^9 m^3 in 1975. Thus, the average consumptive use of water by all plants was computed to be 1.6 l/kWh. Using the previously cited estimate of 2.5 l/kWh for closed-cycle cooling, the unit consumptive use by once-through cooling systems was computed to be 1.0 l/kWh; annual water consumption was calculated to be 1.9×10^6 m^3 by closed-cycle systems and 1.2×10^6 m^3 by once-through systems. This suggests that consumptive use by once-through systems is 40 per cent, rather than the 50 per cent generally assumed. Considering the possible errors in the data used, especially in the estimated water consumption, and considering that the calculation makes no distinction between cooling towers and other closed-cycle systems, the discrepancy is considered acceptable.

Accordingly, for purposes of computing world water consumption, an average value of 1.6 l/kWh was used for all fossil-fueled electric plants, on the assumption that United States conditions were close to typical. The proportion of generation by closed-cycle plants has been increasing in recent years and is expected to continue to increase in most countries (OECD, 1981). Thus, the average consumptive-use figure will edge upward for the rest of this century as new generating facilities come on line and old ones are retired.

Nuclear plants of the light-water reactor (LW) type generally operate at about 31 per cent overall thermal efficiency, indicating that 69 per cent of the thermal energy input is wasted. Because nuclear plants discharge no heat with flue gases, it is generally accepted that 95 per cent of the waste heat, or 66 per cent of the total energy input, is dissipated by the cooling system. Thus, a nuclear plant of the same energy output as an oil- or gas-fired plant at 36 per cent thermal efficiency and having a unit-consumptive-use value of 2.3 l/kWh would require proportionally more cooling water in the ratio of:

$$2.3 \text{ l/kWh} \times 1.4 = 3.2 \text{ l/kWh}, \qquad (6)$$

where 1.4 is the ratio of heat dissipated to cooling in a LWR plant to heat dissipated in a comparable fossil-fueled plant. This 3.2 l/kWh value is quite close to that quoted by the U.S. Department of Energy Environmental Information Handbook, 3.0 l/kWh in metric units. This value has been used in Table 3.2A and in calculating world water consumption. Because most nuclear plants rely on closed-system cooling, no adjustment was deemed necessary to account for once-through cooling, as was done in the case of fossil-fueled plants.

Another type of nuclear power plant, one using high-temperature gas-cooled reactors (HTGR), operates at an overall thermal efficiency of about 40 per cent. It, too, has no flue gases, so 95 per cent of the waste heat is assumed to be dissipated through the cooling system. Using a calculation similar to Equation 6 above, the unit consumptive use of such a plant is estimated at 2.2 l/kWh if cooling towers are used. However, only a small part of the world's nuclear generation is by HTGRs, mainly in the United Kingdom. Because the unit-consumptive-use value of 3.0 l/kWh, adopted for calculating world water consumption by nuclear plants, is lower than the 3.2 l/kWh calculated earlier, no additional adjustment is warranted to allow for generation by HTGRs.

Light-water breeder reactors for electric generation are coming into service in Europe and are expected to assume increasing importance in nuclear-power generation over the next two decades. However, this should make no difference with respect to water consumption by the power plant, because the U.S. Department of Energy (1980) estimates their unit water consumption at the same value as standard light-water reactor plants.

Geothermal electrical generation made up only 0.2 per cent of world electric-power generation in 1980, and is not expected to exceed 1 per cent in the next century. However, it is of special interest because it has the highest unit consumptive use of all the processes identified in the energy sector on Table 3.2A. The reason for this high water consumption is the inherently low thermal efficiency of the conversion processes, which must use low-temperature, low-pressure steam an input to turbines. Even the most efficient plants, tapping dry steam at the Geysers Geothermal Field, California, USA, operate at only about 15 per cent overall thermal efficiency. Plants that must rely on separating steam from hot water generally are in the range of 7-10 per cent thermal efficiency. Because the water consumption is so high, this aspect has received considerable attention in the literature.

Detailed information is available from the Geysers Geothermal Field, California, USA, (dry-steam supply, cooling towers), for Wairakei, New Zealand (hot-water supply, once-through cooling), and from Broadlands, New Zealand (hot-water supply, cooling towers).

The U.S. Department of Energy Environmental Information Handbook (1980) presents information for Geysers Unit 12, a 110-MW_e plant operating at 15 per cent overall thermal efficiency and 75 per cent capacity factor. Water-vapor input is shown as 1,960 acre-feet per Btu x 10^{12}, cooling-tower dissipation as 1,640 acre-feet per Btu x 10^{12}, and the difference, 320 acre-feet per Btu x 10^{12} consisting of cooling-tower blowdown and surplus condensate, as injected to the producing zone. In metric units, this indicates unit water consumption of 6.8 l/kWh. This is confirmed in Kestin (1980, p. 948), who indicates unit water consumption of 6.9 l/kWh by Geysers Units 5-10 of 53-MW_e each.

Axtmann (1974) presents information on the Wairakei Power Plant, New Zealand, a 143-MW_e plant operating at 7.5 per cent overall thermal efficiency and genearting 1,233 GWh per year. At Wairakei, a substantial part of the total energy available is discharged directly to the atmosphere from silencers on boreholes and from open trenches, without reaching the power plant. The plant employs once-through cooling, with discharge to the nearby Wairakei River. Axtmann computes water consumption at 6.7 x 10^6 m^3 per year from the silencers and ditches, and 9.8 x 10^6 m^3 per year as increased evaporation from the Wairakei River, indicating unit consumptive use of about 13 l/kWh.

Information supplied to Unesco by the New Zealand National Committee for the IHP includes design data for the Ohaki Power Project, under development at Broadlands, New Zealand. As of 1979, 34 wells had been drilled (of which 16 were productive) in a field of 4 km^2 extent. Most of the wells tap a zone at 500-1,200 m depth, with producing zone temperatures of 240°-280^oC. Design data for the power plant indicate water/steam input at full operation of 4,190 metric tons/hour (t/hr) at 182^oC, or 1,633 MW, of which 2,705 t/hr is rejected by steam separators at 102^oC, or 380 MW. The gross electrical generation is 165 MW, of which 150 MW is delivered to the electric-power system. Water vapor is released to the atmosphere at a rate of 1,214 t/hr, and 270 t/hr of surplus condensate and blowdown is discharged at 30^oC. Reject water and condensate are reinjected. The overall thermal efficiency appears to be about 9 per cent. Unit water consumption is estimated at 1,214 metric tons per 150 MW, or 8.1 l/kWh. The lower water consumption than at Wairakei evidently is due to higher thermal efficiency achieved through more effective recovery of input energy by employing two stages of steam separation, a primary separation and a flash separation.

Information is given by Robertson in Kestin (1980) on an electric power plant under construction at Heber, in the Imperial Valley, California, USA. This 50-MW plant, which is designed for use of brine arriving at the plant at 170^oC, will reject 463 MW_t through cooling towers and will require 12,793 l/hr makeup. The unit consumptive use is calculated to be 15 l/kWh. The overall thermal efficiency is given as about 10 per cent.

Because nearly half (47 per cent) of the world's geothermal generation in 1979 was at the Geysers and most of the rest was by hot-water systems operating at low efficiency, the United States water consumption was computed separately at 6.9 l/kWh. Elsewhere an average value of 15 l/kWh was considered appropriate, because most plants obtained their energy from hot-water reservoirs and dissipated waste heat through cooling towers.

3.1.7. Oil-shale processing

The sources of information on oil-shale mining and processing differ considerably depending on the assumptions involved. Early planning by the U.S. Department of the Interior (1973) calculated unit consumptive use of water of 2,700-4,400 l/toe, with the largest portion accounted for in disposal of retorted shale, which represented 43 per cent of the total water consumption. Brown and others (1977), assuming the most water-efficient design alternatives, including disposing of retorted shale at a much lower water content than was assumed in the U.S. Department of the Interior study, calculated a unit consumptive use for shale mining and processing of 2,500 l/toe. More recently, Probstein and Gold (1978) estimated unit consumptive use in oil-shale processes at 2,700 l/toe for direct retorting and 4,700 l/toe for indirect retorting.

Water requirements for modified in-situ conversion are significantly lower than for full surface processing, because only one-fifth as much ore is removed as in full surface processing, and disposal of processed shale, accordingly, consumes much less water per unit of production. The detailed development plan for Federal lease C-a in Colorado, USA, for example, at a production level of 3.8 million toe per year, indicates consumption of about 5.5 x 10^6 m^3 per year of water, or a unit consumptive use of 1,400 l/toe. (Gulf Oil Corporation/Standard Oil Company, 1977).

Until the various processes reach commercial scale, it will be difficult to determine the true water-consumption rates. However, because there is at present little commercial-scale production, this does not affect computation of current water use in the energy sector. In estimating future water consumption, the uncertainties surrounding unit-consumptive-use values probably are no more significant than the overall uncertainty of projecting energy production. Accordingly, a rough estimate of 3,500 l/toe has been used, reflecting a compromise between the high and low estimates from various sources.

3.1.8. Tar-sands processing

As explained in Section 2.1.4.2, the main consumption of water in tar-sands conversion is for disposal of the sludge that results from the washing of the bituminous sand to extract the organic matter. Water entrained in this sludge over the short term of decades represents a consumptive use, which was estimated by Camp (1976) at 3,000 liters per ton of product. Information provided to Unesco by the Inland Waters Directorate, Environment Canada indicated water consumption of 47 x 10^6 m^3 per year for oil-sands and heavy-oil production; however, because the water consumption attributable to oil-sands production was not identified, the unit consumptive use estimated by Camp can neither be confirmed nor denied. Accordingly, the 3,000 l/toe estimated above is considerd appropriate for world-wide extrapolation.

Canadian tar-sands production represents the only commercial-scale synthetic fuels production for which specific information is available. Therefore, it is shown as a separate entry under synthetic fuels in the data for Canada in Tables 3.1 and 3.2B. Because the amounts of energy production and water consumption are within the order of rounding of the totals for the OECD countries and the world, the synthetic fuels entry has not been carried forward as a separate entry to those larger entities.

3.1.9. Coal-conversion process

As noted earlier, there is a vast array of candidate processes in the fields of coal gasification, coal liquefaction, and production of clean solid fuel. Each has a different unit consumptive use, depending on the process employed, its overall thermal efficiency, and assumptions regarding proportions of waste heat to be dissipated by air cooling and wet cooling, respectively. The world's only large-scale operating project is in South Africa, and information on water use at that complex was not available. A good index to relative water consumption is overall thermal efficiency of the process (Section 2.4.3); generally, the more complex the chemical processing, the lower the thermal efficiency. Probstein and Gold (1978) offer probably the most comprehensive treatment of water use in synthetic fuel conversion and present ranges of estimates based on maximum, intermediate, and minimum levels of wet cooling. In view of the fact that there is negligible use of water in synthetic-fuel conversion to date, it was considered reasonable for the purpose of estimating future water use to use a single unit-consumptive-use value for each basic form of conversion. This was estimated as the median value in the intermediate range given by Probstein and Gold (1978, t. 9-12), as shown in the table below:

Form of conversion	Range, l/toe	Median, l/toe
Coal gasification	1,800-4,300	3,000
Coal liquefaction	1,600-3,000	2,100
Solid fuel production	700-2,200	1,200

3.2 ENERGY PRODUCTION, 1980

The following tables were compiled from the 1980 Yearbook of World Energy Statistics (United Nations, 1982), Energy Balances of the OECD Countries 1976-1980 (OECD, 1982a, and Energy in the Developing Countries (World Bank, 1980). The format generally follows that of the OECD report, but many entries that would be of no interest in the context of this report have been omitted.

Explanatory Notes

- Units. This report follows OECD usage in expressing energy totals in tons of oil equivalent (toe) because it is an easily understood unit and because quantities of oil are an important element in policy decisions. A ton of oil equivalent is defined as 10^7 kcal, a convenient measure, although it is somewhat lower than the average heat content of crude oil. Throughout this report, 1 ton means 1 metric ton of 1,000 kg unless identified otherwise. Electrical generation is shown in the table as gigawatt hours (GWh); 1 gigawatt hour = 1 million kilowatt hours (kWh).

- Column headings. "Solid fuels" includes all solid fuels, both primary and derived, coke-oven gas, blast-furnace gas, and non-commercial fuels (peat, wood, etc.) in certain countries. (See OECD, 1982a, for further detail.) "Oil and LNG" includes crude oil and natural-gas liquids. "Gas" is natural gas excluding natural-gas liquids. In the "Nuclear," Hydro," "Geothermal" columns, the figures shown as indigenous production represent the oil equivalent energy needed to produce the electrical generation shown, assuming a plant efficiency of 34.4 per cent. "Refining" is the output of refineries, in toe. "Electric generation" shows GWh generated by each type. "Total electric" is the sum of items of electric generation.

- Countries by economic groupings. "OECD Countries" are: Canada, the United States, Japan, Australia, New Zealand, Austria, Belgium, Denmark, Finland, France, the Federal Republic of Germany, Greece, Iceland, Ireland, Italy, Luxembourg, the Netherlands, Norway, Portugal, Spain, Sweden, Switzerland, Turkey, and the United Kingdom.

- "Central Planned Economies" are: the USSR, Poland, the German Democratic Republic, Czechoslovakia, Hungary, Bulgaria, the People's Republic of China, Democratic People's Republic of Korea, the Socialist Republic of Vietnam, Lao People's Democratic Republic, Democratic Kampuchea, Yugoslavia, Albania, Mongolia, and Cuba. "Developing Countries" follows World Bank classification and includes all the countries of Asia, Africa, Europe, Latin America, and the Pacific not included in "OECD" or "Central Planned Economies" except several small countries that are not members of the World Bank. "OECD, Europe" includes all members of OECD except Canada, the United States, Japan, Australia, and New Zealand.

Table 3.2. Energy production, 1980.

Note: Quantities are in millions of metric tons oil equivalent (Mtoe) except electrical generation, which is in gigawatt hours (GWh).

		World				
Category	Solid	Oil & LNG	Gas	Nuclear	Hydro	Geothermal
Indigenous production	1,825.17	3,099.01	1,291.99	57.87	150.54	1.13
Refining		2,789				
Electrical generation, GWh	I-------------5,802,661------------I			672,957	1,750,425	13,150
Total electric - 8,239,193 GWh						

Table 3.2. Energy production, 1980--Continued.

OECD Countries

Category	Solid	Oil & LNG	Gas	Nuclear	Hydro	Geothermal
Indigenous production	796.46	711.99	689.21	51.50	92.63	0.84
Refining		1,753.34				
Electrical generation, GWh	2,083,383	901,741	616,634	598,883	1,077,061	9,811

Total electric - 5,286,973 GWh

Central Planned Economies

Category	Solid	Oil & LNG	Gas	Nuclear	Hydro	Geothermal
Indigenous production	956.20	727.98	416.03	7.01	24.51	0
Refining		633.36				
Electrical generation, GWh	------------	1,693,120	------------	81,500	284,949	0

Total electric - 2,059,369 GWh

Developing Countries

Category	Solid	Oil & LNG	Gas	Nuclear	Hydro	Geothermal
Indigenous production	89.54	1,666.14	151.08	1.34	31.97	0.29
Refining		403				
Electrical generation, GWh	------------	454,783	------------	15,541	371,789	3,389

Total electric - 845,435 GWh

OECD, Europe

Category	Solid	Oil & LNG	Gas	Nuclear	Hydro	Geothermal
Indigenous production	225.73	121.94	139.46	18.42	35.85	0.21
Refining		658.62				
Electrical generation, GWh	633,512	352,026	121,604	214,209	416,872	2,385

Total electric - 1,742,178 GWh

United States

Category	Solid	Oil & LNG	Gas	Nuclear	Hydro	Geothermal
Indigenous production	471.11	484.22	455.22	22.89	24.28	0.44
Refining		734.30				
Electrical generation, GWh	1,276,487	270,346	360,665	266,183	282,333	5,073

Total electric - 2,480,937 GWh

Canada

Category	Solid	Oil & LNG	Gas	Nuclear	Hydro	Geothermal
Indigenous production	25.57	82.90	64.11	3.08	21.68	0
Refining		93.51				
Synthetic fuels		6.21				
Electrical generation, GWh	53,982	15,136	11,035	35,880	252,120	0

Total electric - 386,153 GWh

3.3 WATER CONSUMPTION IN THE ENERGY SECTOR, 1980

The following compilation (Table 3.3A) was based on application of values for average unit consumptive use of water in different processes (Figure 3.3A) to the corresponding energy production statistics given in Table 3.2, as explained in Chapter 3.1. The format follows that of Table 3.2 for ease of comparison.

Unweighted unit-consumptive-use values, on which the weighted average values of Table 3.3A are based, are presented in Appendix C. Water consumption in Table 3.3B and shown in Figure 3.3B is given in millions of cubic meters ($m^3 \times 10^6$); quantities have been rounded independently to two significant figures; hence, totals may not compare precisely. As indicated in footnotes to the table, computations of water consumption in the United States and Canada have in some categories been handled differently than in other groupings; therefore, totals for those two countries are shown separately, in Table 3.3B. Note that $1{,}000 m^3/Mtoe = 1$ l/toe, and $1{,}000 m^3/GWh = 1$ l/kWh.

Table 3.3A. Average unit-consumptive-use values.

Note: Amounts are expressed in liters per ton oil equivalent (l/toe), except for electrical generation, which is in liters per kilowatt hour (l/kWh).

Production	l/toe or l/kWh	
Coal mining[1]	100	l/toe
Oil production[2]	1,100	"
Gas processing and transmission	240	"
Refining		
Oil refining	1,200	"
Nuclear fuel cycle[3]	970	"
Steam-electric power generation		
Fossil-fueled plants	1.6	l/kWh
Nuclear plants	3.0	"
Geothermal plants[4]	15.0	"
Oil-shale processing	3,500	l/toe
Tar-sands processing	3,000	"
Coal-conversion processes		
Coal gasification	3,000	"
Coal liquefaction	2,100	"
Solid-fuel production	1,200	"

[1] Unit-consumptive use for the United States computed using consumptive-use value of 150 l/toe to make allowance for beneficiation (Section 3.1).

[2] Canada and the United States computed directly (Section 3.1.2), other countries at 10 per cent of 1,100 l/toe; world and OECD totals adjusted to reflect difference.

[3] Canada computed using unit-consumptive-use value of 1,690 l/toe (Section 3.1.5).

[4] The United States computed using unit-consumptive-use value of 6.9 l/kWh (see Section 3.1.6); world and OECD totals adjusted to reflect the difference.

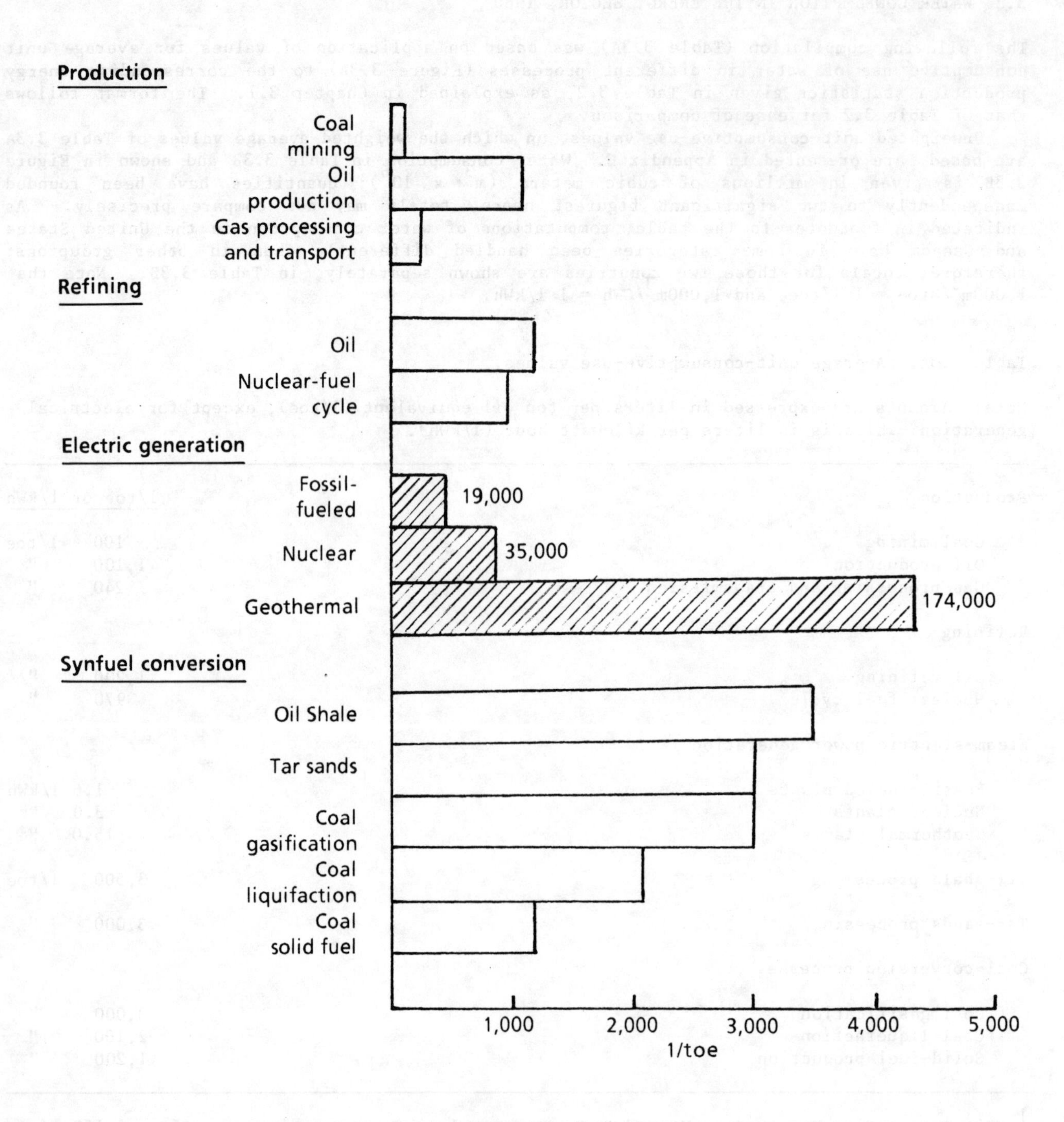

Figure 3.3A. Average unit-consumptive-use values used in calculating water consumption by energy processes. (Patterned bars plotted at 1/40th scale.)

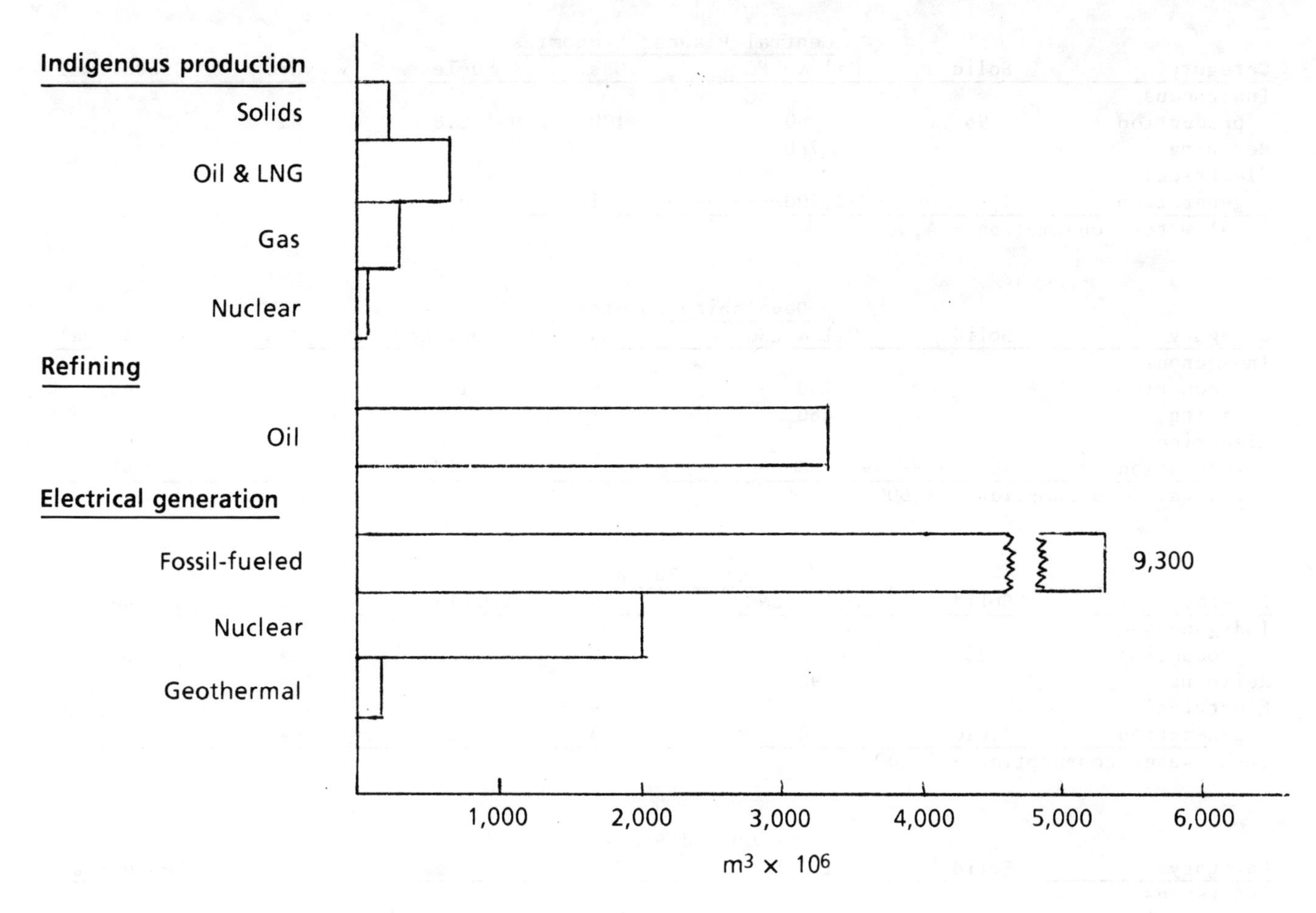

Figure 3.3B. World annual water consumption in the energy sector, 1980.

Table 3.3B. Water consumption in the energy sector, 1980.

Note: Quantities are given in millions of cubic meters, m x 10^6 per year.

			World			
Category	Solid	Oil & LNG	Gas	Nuclear	Hyrdro	Geothermal
Indigenous production	210	650	310	58	na	na
Refining		3,300				
Electrical generation	|-----------9,300--------------|			2,000	na	150
Total water consumption - 16,000						

			OECD Countries			
Category	Solid	Oil & LNG	Gas	Nuclear	Hydro	Geothermal
Indigenous production	100	380	170	52	na	na
Refining		2,100				
Electrical generation	3,300	1,400	990	1,800	na	100
Total water consumption - 10,000						

Table 3.3B Water consumption in the energy sector, 1980--Continued.

Central Planned Economies

Category	Solid	Oil & LNG	Gas	Nuclear	Hydro	Geothermal
Indigneous production	96	80	100	6.8	na	na
Refining		760				
Electrical generation		2,700		240	na	0

Total water consumption - 4,000

Developing Countries

Category	Solid	Oil & LNG	Gas	Nuclear	Hydro	Geothermal
Indigenous production	9	180	36	1.3	na	na
Refining		480				
Electrical generation		780		47	na	51

Total water consumption - 1,600

OECD, Europe

Category	Solid	Oil & LNG	Gas	Nuclear	Hydro	Geothermal
Indigenous production	23	13	38	18	na	na
Refining		790				
Electrical generation	1,010	560	190	640	na	35

Total water consumption - 3,300

United States

Category	Solid	Oil & LNG	Gas	Nulcear	Hydro	Geothermal
Indigenous production	71	300	110	22	na	na
Refining		880				
Electrical generation	2,000	430	580	800	na	30

Total water consumption - 5,200

Canada

Category	Solid	Oil & LNG	Gas	Nuclear	Hydro	Geothermal
Indigenous production	2.6	34	17	5.2	na	na
Refining		110				
Synthetic fuels		18				
Electrical generation	86	24	11	110	na	0

Total water consumption - 410

4 Future energy use and related water consumption through the year 2000

Since 1973, numerous forecasters have projected energy supply and consumption into the future, and most of these estimates have soon been discarded, mainly because world oil-price levels and economic growth rates have departed radically from the underlying assumptions of the forecasts. For the purpose of this report, projections published by the International Energy Agency (IEA) in "World Energy Outlook" (OECD, 1982b) have been used.

IEA developed detailed economic models based on the reaction of energy prices and demand to sharp rises in oil prices in 1973 and 1979. Two detailed models were developed: (1) a high-demand scenario assuming constant oil price and high economic growth and (2) a more realistic, low-demand scenario assuming rising oil price and low economic growth. Application of the second model to crude oil prices since 1973 accurately predicted demand for 1980 and 1981, and that low demand is related to the present (1983) world oil glut, low economic growth rates, and high unemployment in the industrial nations.

These quantitative models illustrate what could happen throughout the rest of the century under prevailing conditions if energy demand and supply were merely functions of variations in price and economic growth. The scenarios differ in numerical results, but both point to a future situation of undersupply in which market equilibrium might be restored only through precipitous oil price rises at the cost of reduced economic growth. In fact, these scenarios differ significantly from a scenario that would result from meeting energy policy objectives to which the OECD countries are already committed. The basic policy is to increase efficiency of energy use and to replace oil through promoting oil substitution and rapid expansion in the use of coal, natural gas, nuclear power, and other available sources of energy.

On the basis of these objectives, the IEA adopted a "reference scenario" for projecting oil demand to the year 2,000 which embodies structural changes in energy demand due to policy actions. Table 4.0 compares key quantities of 1980 actual conditions with the "reference scenarios" for 1990 and 2000.

Table 4. Comparison of 1980 OECD energy requirements with reference scenarios

	1980	1990	2000
		(Mtoe except as noted)	
Total primary energy (TPE) requirement	3,812	4,596	5,502
Total final consumption	2,670	3,108	3,581
Fuel requirements			
Coal	812	1,203	1,854
Natural gas	735	898	1,016
Nuclear	145	412	644
Oil and LNG	1,793	1,741	1,499
Net oil imports	1,180	1,060	816
Energy efficiency			
TPE/GDP ratio (1973 = 100)	89.7	78.3	68.4
Gross Domestic Product (GDP) (billion 1980 US $)	7,543	10,200	13,900

The IEA projections indicate declining real prices for oil from 1980 to 1985, followed by rising real prices over the period 1985-2000. If these projections are realized, net oil imports to the OECD countries are predicted to decline by 45 per cent from 1980 to 2000 while substitute fuels, principally coal and nuclear, show large gains. The shift in fuels is accompanied by a decline in the TPE/GDP ratio and near doubling of GDP in constant dollars.

The rapid increase in the real price of oil since 1973 has increased interest in a major shift away from fossil fuels in general, a shift that may eventually lead to the world's obtaining a major part of its energy from abundant renewable resources, such as nuclear fusion and solar energy. However, unless unexpected technological breakthroughs occur, large-scale use of such resources will not occur in this century. The key events expected for the next two decades are:

1. Economic growth in the industrialized countries at a real rate of about 3 per cent annually from 1980 to 2000, compared with 4 per cent annually from 1965 to 1973.

2. Real energy costs declining from 1980 to 1985, then rising through 2000 because of a limited supply of conventional oil and the high cost of alternative fuels.

3. World energy demand growing at about 2.9 per cent annually through 2000.

4. Only a modest increase in world oil supply and an actual decrease in oil imports to the OECD countries through improvements in energy efficiency and substitution of other fuels.

5. Most energy use, where consumers have a choice, involving greater use of coal and nuclear energy.

6. Use of oil concentrated in special high-value markets, such as transportation, where substitution is unlikely over the next two decades.

7. No large-scale production of synthetic fuels before 2000, barring unforeseeable supply interruptions or major wars.

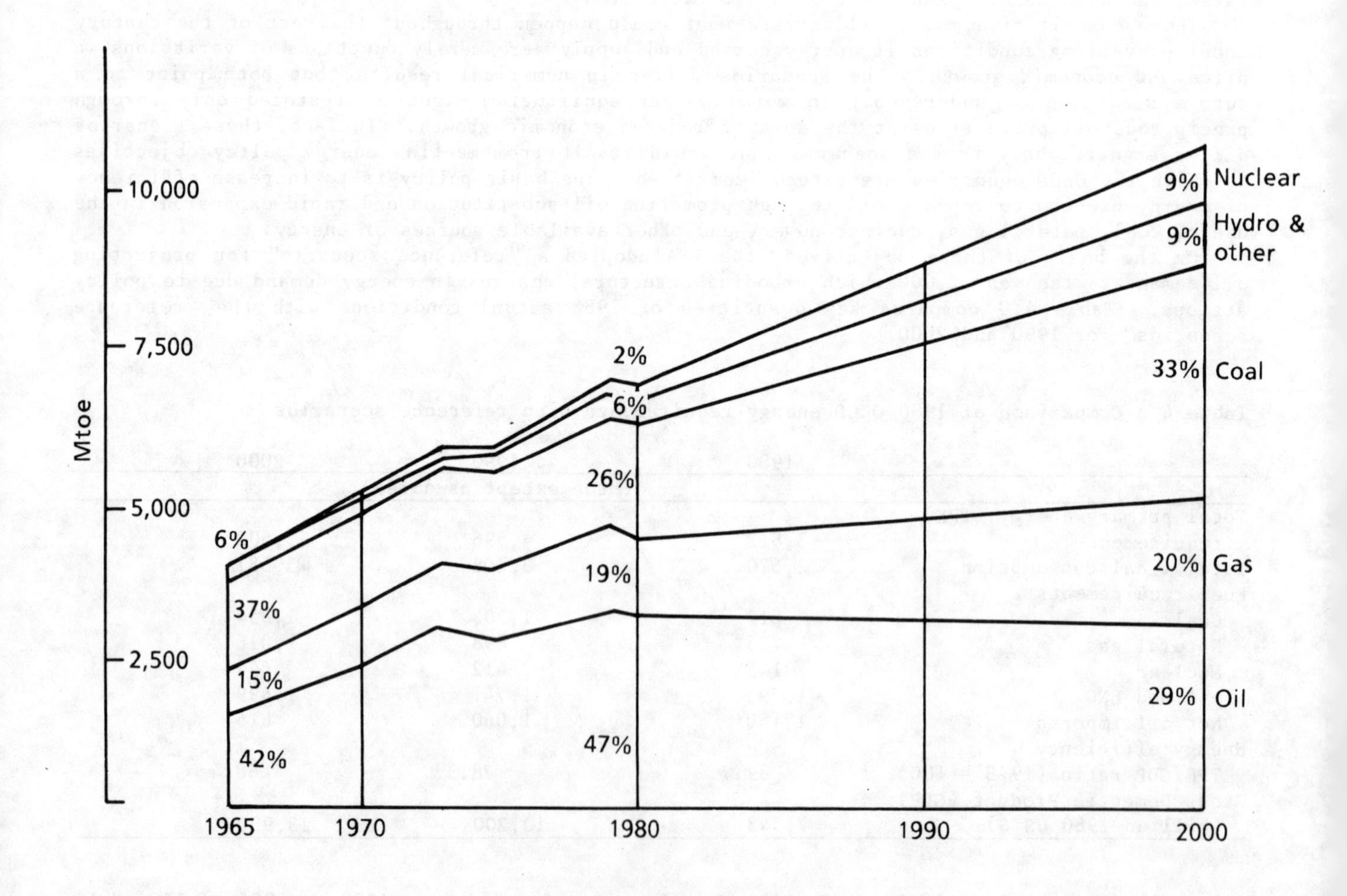

Figure 4. Graph of world energy supply, 1965-2000.

World economic growth rates over the period 1965-1973 ranged from a little more than 4 per cent annually in the low-growth developing countries to nearly 8 per cent annually in the oil-exporting nations and the high-growth developing countries. However, the world average approximated the growth in the OECD countries, a little less than 5 per cent per year, and in the Central Planned Economies was about 5.5 per cent per year. Under the assumptions for 1980-2000, growth rates in the oil-exporting countries will range between 4.8 and 7.1 per cent annually. In the non-OPEC developing countries, economic growth will range between 3.5 and 4.8 per cent per year. In the Central Planned Economies, which as a group are energy self-sufficient, economic growth will be about 2.5 per cent per year, and in the OECD countries, 2.7 per cent annually (Figure 4).

The major industrial countries have made significant reductions in energy consumption per unit of Gross Domestic Product over the past 10 years. These reductions have been due to a combination of improvements in energy efficiency, energy conservation measures, and changes in the mix of goods and services. Compared with 1973, the ratio of total energy requirements per unit of GDP was -16 per cent in the OECD countries in 1981. The most significant reductions to date have been in the industrial sector; by 2000, development of more fuel efficient automobiles is expected to be a major contributing factor.

World energy demand is expected to grow by 80 per cent through the end of the century, reflecting changes in the energy/GNP relationship projected above. The major trends are a general flattening of growth of demand in the OECD countries, resulting in a decreased share of world demand in 2000, coupled with stronger energy demand growth in the developing countries and little change in the percentage of demand among the Central Planned Economies.

Trends in world energy supply are shown in Figure 4. The sharp reduction in growth of oil supply is striking, declining from 47 per cent of total energy supply in 1980 to 29 per cent in 2000. Significant increases in the proportion of supply are shown for coal, nuclear, and hydroelectric and other, and little relative change is shown for natural gas. The trends over three time periods are shown below:

World Energy Supply Growth, Per Cent Per Year

Category	1965-1973	1973-1980	1980-2000
Oil	7.7	2.2	0.3
Gas	7.3	3.6	3.1
Coal	1.0	2.4	3.6
Nuclear	27.8	20.9	8.0
Hydroelectric and other	3.9	4.6	3.9
Average	5.3	2.9	2.8

While oil is projected to decline sharply in relative importance, natural gas is expected to maintain its relative share of the market at about 20 per cent. Coal is expected to replace oil and gas in major industrial markets and electric generation and by 2000 should exceed oil as the largest energy commodity. The growth in the nuclear sector, entirely for electric-power generation, is relatively assured because investments are already being made in the power plants represented, although growth of nuclear power has not been as rapid as earlier forecasts predicted. Hydroelectric power growth is expected mainly in South America at a few highly favorable sites. Also included in the "other" category are geothermal, solar, wind power, and tidal power, none of which are expected to make major breakthroughs over the next two decades.

Electric-power generation is projected to grow at a rate of 3.2 per cent through 2000, somewhat faster then energy growth in general. About one-third of the growth in electrical generation is expected to be in the nuclear area. The trend away from oil and gas is most evident, coal, nuclear, and hydroelectric and other making up the difference.

4.1. FUTURE WATER CONSUMPTION

The net effect of growth of energy supply and "World Energy Outlook", OECD (1982b), with respect to water consumption is expected to result in a 137 per cent increase in consumption in the energy sector by 2000. In estimating water consumption, it was first necessary to array energy use in 2000 by supply sectors. This was done mainly by using "World Energy Outlook" in conjunction with the "1980 Yearbook of Energy Statistics" (United Nations, 1982). This information is presented in Table 4.1A of this report.

The next step was to adjust values for unit consumptive use of water to modified conditions as of 2000. The principal changes expected in unit consumptive use are in oil production, oil refining, the nuclear-fuel cycle, and in fossil-fueled electrical generation. Other processes, although shifting in amount and percentage of total supply, are not expected to change greatly in unit consumption of water.

The change in unit consumptive use of water in oil production is upward by a factor of five. This is to accommodate a great increase worldwide in production of oil by secondary- and tertiary-recovery processes. As the rate of new oil discoveries continues its expected decline, there will be strong economic incentive to maintain production by more sophisticated and expensive means. These methods, which currently account for about 40 per cent of United States oil production and perhaps 10 per cent worldwide, are projected herein to make up 50 per cent of world production by 2000.

Unit consumptive use of water in oil refining has been adjusted upward from 1,200 l/toe, currently, to 1,500 l/toe as of 2000. The rationale for this change is the trend toward greater use of recirculation of water and fewer once-through cooling systems, as noted by Evers (1975). These changes, which are directed mainly at improving overall thermal efficiency, nearly always result in increased evaporative demand and, hence, increased consumption of water.

Major changes are expected also in water consumption in the nuclear fuel cycle. The demand for nuclear fuel is projected to show a significant increase by 2000. Some of this demand, in some countries, will be met by reprocessing of fuel from existing reactors. In the reprocessing cycle, plutonium extracted from used fuel elements is routed directly to fuel-fabrication facilities, but salvageable uranium from the primary cycle must return through another enrichment stage. Likewise, in the use of breeder reactors the additional fuel produced must undergo considerable processing. However, it is expected that these developments could reduce the fuel-cycle requirements by about 15 per cent and the enrichment requirement by about 20 per cent under full implementation. With respect to water consumption, the mining and milling requirement would be reduced proportionally, but savings in other phases of the nuclear fuel cycle cannot be estimated at this time because many key development decisions are subject to great uncertainty. Another trend that will lead to reduced water consumption in the nuclear fuel cycle is gradual replacement of gaseous-diffusion enrichment by the more energy-efficient gas-centrifuge method. This results in an immediate savings of 400 l/toe in unit consumptive use, although the main objective is reduced electrical consumption. New enrichment facilities no doubt will employ this more energy efficient process although older plants probably will still be in operation at the end of the century. Reflecting these various expected improvements in efficiency, the unit consumptive use of water in the nuclear fuel cycle has been adjusted downward to 500 l/toe.

The other major change expected in consumptive use of water is in fossil-fueled steam-electric generation, which is projected to increase by about 50 per cent by 2000, almost entirely by greater reliance on coal. Most new plants currently going into service in North America and the OECD countries are being equipped with closed cooling systems. In part this is because of general scaling up of plant size; many new plants are achieving economies of scale with installed capacity of 2,000 MW or more. The cooling demand of such large plants is enormous and exceeds the low flow of most rivers. Moreover, the waste heat discharged from large plants can have severe impacts on the ecosystems of the receiving streams. Thus, closed-cycle cooling, despite higher costs and greater evaporative demand, is usually the design alternative selected. The United States recorded an increase in electricity generated in plants cooled by closed-cycle systems from 40 per cent in 1975 to 45 per cent in 1980, and this trend is expected to continue as new plants come on line and old ones are retired. It is estimated that by the year 2000, 80 per cent of fossil-fueled steam-electric plants in the world will be cooled by closed-cycle systems, principally cooling towers. Accordingly, the unit-consumptive-use value for estimating water consumption in 2000 has been adjusted upward, from 1.6 l/toe to 2.2 l/toe.

Table 4.1A. Projected world annual energy production, 2000.

Note: Quantities are in millions of metric tons oil equivalent (Mtoe); except electrical generation, which is in gigawatt hours (GWh).

Category	Solid	Oil & LNG	Gas	Nuclear	Hydroelectric	Geothermal
Indigenous production	3,700	3,200	2,200	945	------944---------	
Refining		3,200				
Electrical generation, GWh	|------8.6×10^6---------|			3.3×10^6	3.6×10^6	0.1×10^6

Total electric - 15.6×10^6 GWh

Projected unit consumptive use of water in all energy processes as of 2000 is shown in Table 4.1B.

As the penultimate step in estimating future water consumption in the energy sector, energy projections were arrayed in Table 4.1A. Finally, the modified estimated unit-consumptive-use values for 2000 were applied to the various processes; the rounded results are presented in Table 4.1C. The grand total of water consumption in the energy sector is 38,000 x 10^6 m^3, 237 per cent that of 1980. The combined effects of increased production and increases in unit comsumptive use of water are most evident in oil production (up 390 per cent), oil refining (up 45 per cent), fossil-fueled electrical generation (up 200 per cent), and nuclear electrical generation (up 800 per cent).

Once again it is appropriate to note that an indeterminate but significant proportion of water consumption as of 2000 will be of marine and estuarine water and, therefore, will not represent competition with other water users for freshwater resources. Most water consumption associated with indigenous production of energy raw materials is dictated by the geologic occurrence of those commodities and, for the most part, is at inland sites. Oil refining is done largely at seacoast locations and depends heavily on estuarine and sea waters for supply; this is not expected to change significantly in the near future. Siting of electric power plants, as now, will be determined by the distribution of electric demand; accordingly, much of the increased generation will depend on estuarine and marine waters for cooling supplies.

Table 4.1.B. Estimated average unit-consumptive-use values as of 2000

Category	l/toe or l/kWh	Remarks
Production		
Coal mining	100 l/toe	No change.
Oil production	1,100 l/toe	Applied to 40 per cent of North American production in 1979; 10 per per cent of other. Applies to 50 per cent of world production in 2000.
Gas processing and transmission	240 l/toe	No change.
Refining		
Oil refining	1,500 l/toe	Raised to accommodate greater use of recirculation and elimination of some once-through cooling.
Nuclear fuel cycle	500 l/toe	Lowered to reflect fuel recycling and higher efficiency of gas-centrifuge enrichment method.
Steam-electric power generation		
Fossil-fueled plants	2.2 l/kWh	Raised to account for greater use of closed-cycle cooling.
Nuclear plants	3.0 l/kWh	No change.
Geothermal plants	15.0 l/kWh	No change.
Oil-shale processing, Tar-sands processing, Coal conversion processes	3,000 l/toe	Estimated average for all synfuels.

Table 4.1C. Projected world annual water consumption in the energy sector, 2000.

Note: Quantities in millions of cubic meters (m^3 x 10^6).

Category	Solid	Oil & LNG	Gas	Nuclear	Hydroelectric	Geothermal
Indigenous production	370	1,800	530	470	na	na
Refining		4,800				
Electrical generation, GWh	|------19,000------------|			9,900	na	1,500

Total water consumption - 38,000 m^3 x 10^6

5 Effects of energy development on water quality

Energy development embraces a wide variety of activities ranging from mining and other extaction technologies through physical and chemical processing to conversion of one form of energy to another. It is perhaps not surprising that effects on water quality are extremely varied and range from trivial to severe. The most intractable effects are found in the extractive phases, which commonly are widely scattered and involve considerable disturbance of the underground or surface environment. The wide geographic scatter makes control measures expensive and difficult to implement. An example of this problem is water pollution by inorganic constituents mobilized mainly through underground coal mining. Drainage of the workings is essential to permit mining, but the drainage, together with the general alteration of the underground environment, causes "acid drainage," the production of large volumes of water characterized by low pH, high dissolved-solids content, and high content of heavy metals, particularly iron and manganese. The only effective treatment is by adjustment of the pH, which reduces the heavy-metal content but does not alter the total dissolved mineral load. This problem in its various manifestations, is common throughout the major coal fields of the world. No economically practicable solution or alternative exists to the present systems of mining coal. Similarly, electric-power-plant cooling inherently results in concentration of mineral matter in the cooling supply, which, in temperate climates, must be disposed of to streams. Objecting to the practices solves nothing; such problems are among the many handicaps that society must endure until economically feasible alternatives are found.

In the following section, the various energy processes described earlier are examined in terms of their impacts on water quality. In table 5.5, each process is rated in general terms for frequency (or likelihood) of occurrence of problems, the areal scale of these problems, the time frame for alleviation of contamination, the severity of impact on society, and the effectiveness of control measures. This rating scheme is highly subjective and is not intended as a quantitative means for comparing one process with another. Rather, it should be thought of as a device for arraying the problems systematically with a view toward deciding which segments of the energy sector warrant special attention with respect to water quality in the future.

It is all too common in the energy sector for public authorities to stress certain aspects of pollution and to devote excessive attention to those matters to the detriment of other problems which, being less noticeable or more commonplace, get less public attention.

After careful study and analysis of the field using formal comparative analysis as a tool for ranking the environmental impacts of various energy processes, the U.S. Department of Energy Office of Environmental Assessments (1980) concluded that there "probably is no way to design and implement a comparative energy/environment paradigm that adequately allows tradeoffs to be made between real-world energy, environmental, and economic considerations." They further concluded that much of the data needed to model the energy sector adequately cannot be quantified and that past attempts at formal comparative analysis are flawed by incompleteness, lack of commensurability, uncertainty, ignorance, subjectivity, and bias.

5.1 EXTRACTION

The main extractive processes in the energy sector include mining of coal, uranium, and other solids including oil shale and tar sands and the extraction through wells of oil, natural gas, and geothermal fluids. Oil shale, tar sands, and uranium also may be extracted by in-situ methods, in which the raw material is converted to a more mobile form in place. The converted material can then be extracted through wells. Some experimentation has been done on in-situ extraction of coal through underground burning which yields gaseous and fluid products, but the technology is still far from commercial application and therefore is not treated in detail in this report.

5.1.1. Coal mining

Coal mining is subdivided into underground mining, surface mining, and coal beneficiation. The latter process is in a sense a form of refining, but because it generally is carried out at or close to the mine site, it has been treated as an aspect of mining in this report.

5.1.1.1. Underground mining. Water-quality degradation is one of the major environmental concerns associated with coal mining. Adverse impacts result from drainage from active and abondoned mines, from leaching and erosion of mine wastes, and from disruption and contamination of aquifers through mining. The water quality factors of greatest concern are low pH, high dissolved-solids content, and high content of various metals, principally iron and manganese.

Acid drainage from underground workings and from coal and refuse storage piles is particularly severe where the coal seams and overburden are rich in sulfur, commonly in the form of pyrite (iron sulfide). As a result of mining, pyrite is exposed to the action of air and water and oxidizes to form sulfuric acid. The acidity causes the solution of compounds of iron and manganese, which impart a foul taste, and of heavy metals such as arsenic, cadmium, chromium, lead, mercury, and nickel, which are toxic to varying degree. The acidity and heavy metals can destroy aquatic life and, together with the objectionable taste due to iron, manganese, and other dissolved mineral matter, can render water unfit for domestic use. In the Appalachian coal fields of the United States, for example, some 16,000 km of streams have been severely degraded through such pollution (Appalachian Regional Commission, 1969). It is estimated that three-fourths of that pollution can be attributed to abandoned underground mines (Johnson and Miller, 1979).

Aquifer disruption is another source of water-quality degradation. Water can be degraded through interconnection of vertically adjacent aquifers and by seepage of contaminants from waste piles. Adjacent aquifers commonly are contaminated through interflow through cracks and fissures resulting from subsidence, or by interception of aquifers during mining. Such effects are highly localized and are characterized by extreme variability (Hollyday and McKenzie, 1973). Ground waters may be contaminated directly by surface streams when cracking extends to land surface. Indeed, streams sometimes disappear completely into mines to reappear somewhere else at lower elevation, generally with impaired water quality. This effect, of course, disrupts both the ground-water and surface-water regimes (Subitsky, 1976). Subsidence, which is the basic cause of the most severe aquifer disruption, is virtually unavoidable in most coal underground mining (Figure 5.1.1.1). With complete-recovery mining techniques, subsidence occurs as mining proceeds and the adverse impacts, although often severe, generally are short-lived. With room-and-pillar mining techniques, however, many adverse effects are drawn out over decades as the coal pillars slowly fail owing to the load of overlying sediments.

5.1.1.2. Surface mining. The most severe impacts from surface mining stem from the disturbance of large areas during mining and from haul roads needed to remove the coal. In mountainous areas with high rainfall, it is not uncommon for sediment yields of streams to rise during mining to 100 times or more the pre-mining sediment yield. High sediment content can destroy aquatic life, fill reservoirs, and bring on severe downstream flooding, thereby endangering human life as well. Although generally not as severe a problem as in the case of undergound mining, surface mining exposes great quantities of fresh mineral matter to the effects of weathering, and acid drainage and its adverse chemical reactions may occur. Where auger mining is carried out as the final phase of surface mining, the horizontal holes act as lateral wells and thus may aggravate an already adverse chemical condition.

5.1.1.3. Coal beneficiation. This is done to upgrade the quality of mine-run coal by removal of inorganic matter such as shale, clay, and pyritic sulfur, so as to meet a customer's specifications. While the objective is to produce a cleaner coal, the process often leads to local pollution of two major types, leaching of waste piles and release of "black water" containing coal and rock fines in suspension from process waste waters. Coal is shipped out as a product, but generally the waste is disposed of in piles at the plant site or in a tailings pond. Because the waste consists largely of fresh mineral matter, it is subject to rapid oxidation and weathering. Leachate from waste piles and tailings ponds is similar chemically to acid drain water and can severely degrade receiving streams and ground waters.

Figure 5.1.1.1. Coal miner setting support at underground mine face under "roll" in mine roof. When such supports are removed, overlying strata cave into mined-out space. (Photo by U.S. Bureau of Mines.)

Discharge of "black water," where tolerated, can be a major nuisance to downstream water users. The problems cited are symptoms of poor design and management, however, and are amenable to simple corrective measures.

5.1.2. Oil and gas extraction

Oil and natural gas are produced through wells from permeable rocks that normally contain saline water in association with the lighter hydrocarbons. Accordingly, considerable saline formation water is produced along with most oil; only negligible amounts of saline water are produced with gas. The handling and disposal of these waste brines constitutes the major pollution problem in oil and gas operations.

5.1.2.1. Primary extraction. This term covers the oil produced from the time a well is completed until production by natural reservoir pressure or pumping is no longer economically viable. Fairly early in their lives, most oil wells begin to produce increasing proportions of formation water, often rising to 90 per cent of the total fluid production. This saline water has little value, and its handling and disposal represent a major cost to the operator. Because high salinity is corrosive to metal, well casings, pipes, and tanks require careful maintenance to prevent the escape of saline waters to the environment. Commonly, such waters are disposed of by injection into non-producing formations.

5.1.2.2. Secondary recovery. This term refers to the process of injecting natural gas, air, or water into the producing zone through injection wells to displace oil and drive it toward producing wells. Secondary recovery offers an ideal solution to the problem of disposal of formation waste water. Being from the same formation, formation waters normally are in chemical equilibrium with the sediments and, therefore, are less likely to cause chemical reactions that could interfere with the injection process. Where produced waters are inadequate in amount, they generally are blended with an available fresh or marine water. However, the foreign waters generally require considerable treatment prior to injection. In secondary recovery, handling of saline waters remains the major potential pollution problem; moreover, it is made somewhat more complex because the injection generally is done under high pressure. Weak, corroded pipes and casings are even more subject to failure, and in the event of a leak in casing opposite a fresh-water zone, the injection water can escape rapidly. Another source of pollution of shallow ground waters is leaky holding ponds, which can infiltrate saline water directly to shallow aquifers.

5.1.2.3. Tertiary recovery. When heat or solvents are used to mobilize oil remaining in a formation, the process is termed "tertiary recovery." Steam is the vehicle usually used to transmit heat to the producing zone. In this technique, steam is generated in surface boilers and is injected as described in Section 2.1.2.3. The produced fluid is a blend of oil and oily water. As in the earlier stages of recovery, handling of formation fluids is the chief threat to water quality, and it is further complicated by the need to deal with hot saline waters, which are even more corrosive than saline waters of normal temperature. A variation of thermal methods involves igniting the oil in the formation, producing a burning front that mobilizes the remaining oil and drives it toward production wells. Formation waters are largely converted to steam, and the production and attendent pollution hazards are similar to those of steam flooding.

The other basic principle employed in tertiary recovery involves the injection into the formation of carbon dioxide, carbonated water, or surfactant solutions to separate residual oil from the sediment grains and displace it toward production wells. Carbon dioxide in solution is chemically active and vigorously attacks steel, thus increasing the leakage hazard. Although carbon dioxide is not objectionable as a contaminant, the carrier water may be. Surfactants used in tertiary recovery are generally organic-chemical solvents, which constitute a pollutant themselves quite aside from the chemical character of the carrier fluid.

To summarize, the contamination hazard to surface and ground waters in oil extraction tends to increase with increasing complexity of the recovery process. Offsetting this is the fact that oil is an expensive commodity and its loss represents an economic loss over and above any pollution damage. Hence, there are strong economic incentives to avoid pollution. Because of the vast number of oil wells throughout the world and their wide dispersal, accidental leaks will occur sporadically, but the legal and financial incentives to avoid pollution are very persuasive.

5.1.2.4. Blowouts. A relatively rare water pollution hazard in oil operations, yet one that can be extremely serious is that of blowouts, or uncontrolled release of oil and gas from wells. This type of accident most often occurs during the drilling of exploration wells where subsurface conditions are poorly known. If a high-pressure zone is encountered unexpectedly, the drilling fluid may be expelled from the borehole by high gas pressure, followed by a massive outflow of oil and gas from the borehole. Commonly the drilling rig and pressure control equipment are destroyed in the ensuing fire, and weeks to months may be required to regain control of the unchecked flow. When blowouts occur on land, the hazard to streams from oil pollution generally is quickly alleviated through construction of dikes to salvage the produced oil. In offshore operations, however, this simple control measure is impossible, and oil discharging directly to the sea floor spreads with the winds and currents over broad areas. The Ixtoc 1 well off the Yucatan Peninsula of Mexico, for example, which blew out on June 3, 1979, reportedly discharged some 600,000 tons of oil before it was brought under control about a year later. The degree of damage from such events is closely related to the qualities of the accident site and varies with the location of the blowout with respect to important marine resources and the direction and strength of currents and winds.

5.1.3. Oil-shale extraction and processing

Water-quality problems associated with oil-shale extraction and processing currently are negligible on a world scale because there is little commercial-scale production. However, should world oil prices recover from their depressed 1983 level, a number of operations currently suspended in Colorado, USA, could bring oil-shale production to the commercial stage in the USA within 10 years. Impacts on water quality are possible in several phases of operations, but the most significant concerns relate to the potential for escape of noxious organic and inorganic compounds from the mining and processing area. Several modes of operation are contemplated. Where ore-grade deposits are found near the surface, they will be mined by open-pit methods and the shale will be processed in surface-retorting facilities. Where rich deposits are economically accessible by undergroung methods, room-and-pillar mining will be used to recover ore from high-grade seams. Where surface or underground mining are impracticable, in-situ techniques will be employed, in which part of the resource is burned in place, the heat retorting a large mass of rock, thus producing liquids and off-gases that can be collected and used. The principal technique of this class is the modified in-situ method, in which about 20 per cent of the ore in a large underground volume is extracted, the ore remaining underground is shattered with explosives, the shattered ore is ignited from the tip, and the burn proceeds downward. Liquids and off-gases moving downward ahead of the burn front are collected from the bottom of the rubble zone. The ore previously extracted is treated in a surface retorting facility.

The main concerns relating to water pollution center on disposal of waste rock. Tremendous tonnages of waste result from a process that extracts no more than 14 per cent of product from each ton of ore. The high-carbonate-content shale is calcined in the high-temperature retorting process, and the waste product that results is somewhat like a very poor grade of cement containing a large fraction of organic compounds. Depending on the retorting process employed, the waste material may be a fine black powder or discrete chunks on the order of 50-75 mm in diameter. Current plans for waste disposal in Colorado, USA, aim at placing the material in dump piles, commonly filling small canyons. The water content will range from about 10 per cent to 25 per cent, and the waste should react with water to form a weakly coherent, concrete-like rock. If subjected to leaching, vast quantities of mineral matter would dissolve and be carried off in the leachate. However, the main areas of production in the United States are quite arid, with mean annual precipitation of 250 mm or less. Under this rain-fall regime, direct infiltration of precipitation from the surface is negligible. Consequently, if runoff is directed away from the waste-disposal areas, the hazard of stream contamination from the waste will be minimal. Table 5.1.3., from the U.S. Department of Energy (1981a), illustrates the wide spectrum of potential pollutants that must be dealt with in oil-shale development.

In surface mining, the principal impact from mining per se is the increased potential sediment transport due to excavation of extensive land areas and construction of access and haul roads. In this respect, surface oil-shale mining is similar to surface coal mining, and careful attention will be needed to minimize the sediment-transport problem.

Underground mining may intercept water-bearing fractures, thus leading to mine drainage of waters that have had an opportunity to leach the natural ore material. Experience in the western United States over more than 40 years in experimental mining of deposits above the zone of saturation suggests few water problems. The ore is dense and tough and, unlike coal, does not fracture readily. Although small seeps are observed, mine drainage has not been a significant issue.

Operations in the zone of saturation, either by underground mining or by in-situ recovery, involve a different range of water problems. Certain zones in the oil-shale sequence are highly permeable and will require massive dewatering efforts; also, much of the water in the oil shale is highly saline owing to solution of salts that are interbedded with

Table 5.1.3. Sources and nature of water effluents from oil-shale facilities.

Source	Pollutants
Direct Sources	
Mine/retort dewatering	Sodium, chemical oxygen demand, carbon disulfide, fluoride, boron, sulfide.
Retorting	
Water separated from crude shale oil	Ammonia carbonate, sodium, sulfate, SCN, chloride, dissolved or suspended organics (e.g., phenolics, amines, organic acids, hydrocarbons, mercaptans, smaller quantities of calcium, magnesium, sulfide, trace elements (e.g., mercury, selenium, arsenic), suspended shale fines.
Condensate from retort gases	Ammonia, carbonate; traces of organic substances and sulfur containing compounds.
Upgrading	
Oily cooling water Process waste water Spent caustic streams Oil-free waste waters	Ammonia, bicarbonate, sulfide, phenols, total dissolved solids, oil and grease.
Cooling water and boiler water blowdowns	Dissolved solids, corrosion controls (e.g., hexavalent chromium).
Raw water treatment	
Chemical sludges Backwash water Blowdown from zeolite softeners	High hardness and dissolved salts.
Sanitary waste water	
Domestic sewage	Human excrement, paper, soap, dirt, food wastes, unstable organic matter, microorganisms.
Indirect Sources	
Leachate from retorted shale	
Leachate and/or runoff	Sodium, magnesium, sulfate, chloride, fluoride, small quantities of organic substances and trace elements.
Runoff and erosion from construction and site use	Sediment, dissolved solids.
Mining and transport	Sediment, dissolved solids.

the shale in places. Saline water produced in dewatering operations may be of value for process water supply; otherwise it will require careful handling and disposal to avoid stream pollution.

Where in-situ retorting is practiced, the shale must be dewatered to permit underground combustion. The waste water produced in this way is a variable quality and must be handled carefully to avoid pollution of surface streams. Following completion of retorting, the water table will rise into the zone of burnt shale, and organic and inorganic compounds will be subject to solution in the ground water. Experience with leaching of surface-retorted shale suggests that the water saturating underground retorts will take large amounts of mineral matter into solution. Ultimately, such ground waters can be expected to discharge at the land surface, where they may pose pollution problems. Large-scale experience with the impact of the in-situ method on ground waters, however, is lacking, and ground-water quality effects may not prove to be a serious problem. Should the opposite prove to be the case, the long-term solution may be to intercept degraded ground waters before they can discharge to streams and to dispose of them elsewhere.

5.1.4. Tar-sands extraction and processing

The extraction of bitumen from tar sands is carried out on a commercial scale only in Alberta, Canada, where two operations currently produce about 17,000 tons per day of crude product. The two operations are similar in that the material is mined by open-pit methods and the hydrocarbons are extracted in a hot-caustic-soda washing process. The extracted bitumen is then treated on-site in a thermal-cracking process similar to the initial stages of oil refining. As in oil refining, there is some effect on stream quality due to controlled release of waste waters and minor accidental releases of organics in the plant area. However, the greatest threat to stream quality is from accidental release of fine-grained wastes that are stored behind high dikes. With an average ore content of only 12 per cent hydrocarbons, tremendous tonnages of waste are generated. Although the raw material is termed "tar sand," it generally contains much clay-sized mineral matter, which remains in suspension to form large volumes of sludge. Because this sludge loses water only very gradually, waste disposal consists of semi-permanent impoundment of the sludge behind sand dikes up to 100 m high. This method of storage is similar to that employed in storage of phosphate-plant wastes in Florida, USA, where failure of retaining dikes has, on occasion, resulted in great downstream damage from the fine sediment.

An estimated 10×10^9 tons of tar sands in Alberta, Canada, are estimated as recoverable by in-situ techniques. A number of large-scale experiments have been carried out using steam-flooding methods similar to those used in tertiary recovery of oil. It is expected that impacts on water resources will be quite similar to those of tertiary-recovery processes in oil production. The main concern centers on the potential for escape of process steam and hydrocarbons from the producing zone to overlying water-bearing zones. This has not proven to be a serious problem in tertiary oil production to date and should not be a serious problem in revovery of bitumen from tar sand where good engineering practice is followed.

5.1.5 Geothermal extraction

The two principal forms of geothermal occurence, vapor-dominated (or dry-steam) resources and water-dominated (or hot-water) resources, have entirely different impacts on the water quality of streams and ground waters and therefore are discussed separately. In the former, wells tapping the geothermal reservoir produce steam containing various proportions of gaseous contaminants, but in the latter, the hot waters are characterized by a high content of dissolved mineral matter, including a relatively high content of toxic metals such as mercury, arsenic, and selenium.

5.1.5.1. Vapor-dominated systems. In the vapor-dominated systems, as represented by the Geysers Field, California, USA, and the Larderallo Field, Italy, gases mixed with steam make up less than 1 per cent and 30 per cent of the well discharge by weight, respectively. The gases consist mainly of carbon dioxide, with lesser amounts of ammonia, hydrogen sulfide, and methane and traces of radon, mercury, and boron. Most of the gas is ejected to the atmosphere before the steam reaches the electric plant, but some gas remaining with the steam is dissolved in the power-plant condensate in the form of ammonium, bicarbonate, and sulfide ions. The presence of these dissolved gases results in a low pH, and these hot solutions are extremely corrosive to metals, concrete, and even wood. Boron is toxic to plants at levels higher than about 1 mg/l, and that is an important consideration in waste disposal if downstream waters are used for irrigation, as is the case at the Geysers Field. Analyses of condensate from the Geysers Field indicate total dissolved solids of about 1,000 mg/l consisting mainly of ammonium and bicarbonate ions. Nevertheless, this water exceeds allowable limits for boron and ammonia for direct discharge to streams. Consequently, the surplus condensate and power-plant blowdown from the Geysers Field have been reinjected into the producing formation since 1969. Because the dissolved-solids content of the waste is low, there has been little problem with maintaining injection rates. Much of the producing area at the Geysers Field is occupied by natural hot springs and steam seeps, so it is difficult to assess whether the reinjected wastes find their way back to the surface. The issue is academic in any event because the numerous natural springs of the area discharge waters to streams that generally are more objectionable than the geothermal field wastes. Ground-water pollution, likewise is not a problem, because no useable aquifers are known in the geothermal field.

5.1.5.2. Water-dominated systems. By way of contrast, water-dominated, or hot-water, systems are characterized by total dissolved solids generally in the range of 2,000 to 20,000 mg/l but reaching as much as 250,000 mg/l at the Niland Field, Imperial Valley, California, USA. Hot saline waters are very effective solvents; accordingly, geothermal waters commonly contain high levels of silica in addition to the common ions Ca, Mg, Na, HCO_3, SO_4, and Cl. Moreover, toxic metals including mercury, arsenic, and selenium are also found in relatively high, harmful concentrations, in most geothermal waters. The volumes of waste liquids from geothermal plants are very large and only two basic methods of disposal are available, direct discharge to streams or the sea and subsurface injection. Both methods are used at developments throughout the world. Historically, geothermal producers have preferred direct discharge because of low cost, but the downstream effects have brought about more stringent controls for environmental, health, and safety reasons.

Outside the United States, direct discharge is widely used, with varying results. At Wairakei, New Zealand, waste fluids have been discharged to the Waikato River for many years. The geothermal fluids represent about 1 per cent of the average flow and have caused some problems downstream. Axtmann (1974) calculates that the geothermal waste discharge from the power plant totals 158 metric tons per year of arsenic in the river water. Arsenic from the river water is concentrated in sediments and weeds to such an extent that it is unhealthy for animals to feed on the weeds. Thermal waste apears to have caused rapid growth of water vegetation and to have impaired the waters for trout. It was to avoid additional stress on the Waikato River that the design for the new Otaki geothermal power development at Broadlands specifies injection of waste geothermal fluids (Gibson, 1979).

At many places, reinjection is impracticable because of the high mineral content of the geothermal water, which tends to precipitate the less soluble constituents in injection wells after the heat has been extracted, thus plugging the wells. Treatment of such spent fluid to remove troublesome mineral constituents will produce large volumes of mineral sludge, which in turn will require careful disposal to avoid contamination of surface and ground waters.

5.1.5.3. Geopressured systems. In the United States, there has been considerable interest in the potential for development of "geopressured brines" of the Gulf of Mexico coast. Wells drilled for oil and gas in coastal areas of Texas and Louisiana commonly penetrate zones in the late Tertiary sequence that contain warm saline water under high pressure, sometimes approaching the pressure caused by the total load of overlying rocks and water. This water, which contains 1,000-10,000 mg/l of dissolved solids, also contains large amounts of methane in solution, which would have high value if the brines could be produced and disposed of at reasonable cost. The disposal of large quantities of saline water without causing surface- or ground-water pollution has been one of the major barriers to the development of the resource. Various schemes have been proposed for extracting the methane and thermal energy from the fluid and using the high pressure to generate electric power; however, recent drilling tests have not been encouraging, and interest in further development is lagging (Kerr, 1980). Because the method is not carried out at commercial scale and does not appear promising, geopressured development is not treated elsewhere in this report.

5.1.6. Uranium mining and milling

Commercial concentrations of uranium ores occur in two principal modes: as primary ores in vein deposits in crystalline igneous rocks and as secondary concentrations precipitated from percolating ground waters in permeable sandy deposits. The latter type is the most common in the United States; most of the rich ore bodies in Canada and southern Africa are of the vein type. The vein-type deposits are found in impermeable crystalline rocks, and their mining presents no special water quality problems. The sedimentary ores, conversely, occur most commonly in aquifer materials, and much dewatering is required to permit mining. The water pumped from permeable ore zones commonly contains uranium and radium in concentrations that preclude direct disposal to streams or ground waters on public-health grounds. The usual approach to disposal under these circumstances is to evaporate the water in large, shallow basins. If the basins leak or fail, the radioactive waters can contaminate surface- and ground-water supplies. This problem must be faced in most underground mining of sedimentary ores and in surface mines that extend below the water table.

Uranium milling, which generally is taken to include on-site chemical concentration processes, poses even more serious water-quality issues. After crushing to sand size, the ore is subjected to an acid-leach process to oxidize and dissolve the uranium minerals. The uranium is extracted in the form of uranyl sulfate, washed, dried, and shipped in the form of uranium oxide to refining plants. Because uranium generally makes up less than 0.1 per cent of the ore, large volumes of waste remain after the uranium is removed. This waste rock material, together with fluid wastes from the wet-separation process, is generally disposed of by slurrying it to piles, where most of the water is decanted and returned to the mill.

As described by Markos and Bush (1980), uranium-mill tailings comprise a residue of ore-forming materials, the silicate matrix of the host rock, and chemicals used in the extraction process. Most of the uranium and vanadium in the ore is removed in the concentration process, but many toxic and radioactive elements remain, among them thorium, radium, arsenic, barium, copper, molybdenum, lead, selenium, and zinc. The silicate matrix consists mainly of quartz, with lesser amounts of feldspar and clay minerals. In the mill tailings, sulfate and chloride from the acid-leach process are the dominant anions. Most of the mobile elements occur in the interstitial space of a sandy or clayey matrix.

The acid-leach process involves two principal reactions. First, the tetravalent uranium in uraninite and other uranium minerals are oxidized, in a two-step process, to uranyl ions. The most commonly used oxidant is sodium chlorate. Second, sulfuric acid is used, resulting in the presence in the solution of sulfate, bisulfate, and hydrogen ions. The sulfate complexes with uranyl ions to form various forms of uranyl sulfates. The tailings commonly contain as much as 1.5 per cent chloride and 6 per cent sulfate by weight. In the acid environment, the carbonates of calcium and magnesium break down and many of the aluminosilicates of the clay minerals dissolve.

The interstitial fluids of the tailings are a concentrated solution high in sulfate and chloride. Chemical disequilibrium exists between the solution and the solid phases as well as among the different components of the solid phase. As a result, the various components continue for many years after emplacement of the tailings to react to attain equilibrium. Various precipitation reactions form readily soluble salts of sulfate and chloride with the major cations and trace metals.

If the tailings are placed in such a way that they are permanently isolated from the hydrologic environment, they do not pose a threat to water quality; however, such isolation is difficult to achieve and maintain. The U.S. Department of Energy is currently carrying out a program to provide for better control of uranium-mill tailings at sites where mills are no longer operating. It is estimated (U.S. Deptartment of Energy, 1981a) that some 100 million metric tons of such waste are located at 20 locations throughout the United States. Emissions of radon-222 are considered the most likely cause of radiation exposure, through inhalation. Plans for eliminating the radioactive exposure hazard include covering the tailings with soil or more impermeable capping and, if appropriate, physically moving the tailings away from populated areas.

The in-situ solution-mining technology for extracting uranium avoids the problem of tailings disposal. In this technique, acid solutions are injected into uranium-bearing aquifer zones; the acid solutions carrying uranium in solution are produced from wells and the uranium is stripped from the solution in an on-site processing plant. The injection solution is then refortified and recirculated. In the usual operation, the producing wells are on the outer perimeter of the area of operation, and they are operated at a higher rate than the injection wells so that the acid solutions will not escape the area. In this way, a small surplus of water is continually generated, which is treated to remove objectionable contaminants and then is released to the environment.

5.2. TRANSPORTATION

The prinicpal forms of transport of energy fuels are trucks, railroads, barges, marine tankers, and pipelines. Truck and rail transport are limited generally to movement of solids, which represent minimal hazards to water resources and therefore are not discussed herein. Of greater interest with respect to water quality effects are the fluid-transport systems, chiefly pipelines and marine tankers. Pipelines are used in oil production to gather crude oil from the individual producing wells and to move it from central gathering points to refineries or marine or river terminals for shipment to distant refineries.

5.2.1. Oil pipelines

The principal problem leading to water contamination is pipeline breaks. Among the hundreds of thousands of kilometers of oil pipelines throughout the world, some accidental spills are statistically inevitable. Oil released in these episodes is highly damaging to aquatic life and can impair the use of streams for other purposes until the oil is cleaned up. Leaks in underground lines generally are localized, but the problem of removing oil from an aquifer is expensive, lengthy, and technically difficult. Most pipelines are buried to a depth of 1 m or more and thus are protected from most sources of surface damage; however, they are subject to damage from excavating equipment.

Loss of oil due to pipeline breaks is minimized in large systems by pressure-sensing devices that shut off pumps when a loss of pressure occurs; this, together with self-actuating valves and baffles, helps to prevent gravity drainage over long distances. With crude oil having a value (1981) of close to US$ 0.25 per liter, even the most profligate operator can ill afford spills, and the high price encourages vigorous efforts to salvage spilled oil. Probably the most serious single spill due to a pipeline break recorded was that of October 15, 1967, when a submarine pipeline off West Delta, Louisiana, USA was severed by the dragging anchor of a ship. In this spill 22,000 metric tons of oil escaped to the littoral environment.

One of the world's largest oil pipelines, and one that has suffered several spills, is the Trans-Alaska Oil Pipeline, which carries some 100 x 10^6 metric tons of oil per year from Prudhoe Bay on the Arctic coast of Alaska 1,000 km south to Valdez for shipment by marine tankers. The pipe, about 1.2 m in diameter, carries warm oil through an area characterized by the occurrence of permafrost over much of the route. Consequently, in the permafrost region the line was built largely above ground, supported on pillars. Where it is buried it is subject to shifting and settlement because of melting and heaving of the unstable frozen ground. Major rivers are crossed with bridge structures. Because of its exposed surface position, the pipeline is more subject to surface mishaps than buried lines and on several occasions has been the target of vandalism and sabotage. Service has been interrupted several times because of accidental breaks and sabotage; however, due to the cold climate the extent of damage and its impact on water resources has been small; the spilled oil congeals quickly as it cools and is easily salvaged.

5.2.2. Coal-slurry lines

Coal-slurry pipelines move finely ground coal in a 50 per cent water slurry in much the same way that oil is pumped through a pipe. Only one major line, the Black Mesa Pipeline in Arizona, USA, is currently in operation, but another large system is scheduled to begin construction soon and several others are in the planning stage in the United States and elsewhere. The Black Mesa line (Section 2.2.1) has operated 99 per cent of the time since it went into operation in 1970 and has had no significant impact on the water environment. The line is buried throughout its 440 km length; thus there is little hazard of breakage except through earth movements or by heavy excavating machinery.

The Energy Transporation Systems, Inc. (ETSI) slurry pipeline, scheduled to begin construction soon, will extend 2,200 km from eastern Wyoming to Baton Rouge, Louisiana, on the lower Mississippi River, USA. It is planned to deliver 22.5 metric tons of coal per year through the 95 cm pipe. Pumps to be positioned at 80-250 km intervals will raise the pressure in the line to 68 kg/cm^2, and the pressure will decline to about 7 kg/cm^2 before being boosted again at the next pumping station. Earlier slurry lines were designed with reservoirs along the route to provide for draining the slurry from the line in the event of a service interruption. However, experience with the Black Mesa line has shown that a coal slurry can be restarted without difficulty even after several days of shutdown. Consequently, the ETSI line will be built without reservoirs, thus removing a possible source of contamination of water resources.

It can be seen that the danger of spills from coal-slurry lines is quite small. Furthermore, the danger of ground-water contamination is much less than in an oil pipeline of comparable size. In the event of a leak too small to be detected by pressure sensors, the slurry would act much like drilling mud and would form a filter cake over the site of the leak, thereby plugging it. At worst, only a small volume of carrier water could escape to the ground-water environment.

5.2.3. Oil tankers

A more serious hazard, albeit to marine waters, is that of oil-tanker spills (Figure 5.2.3). The World Almanac (Lane, 1981) records 13 major tanker spills since 1967 in which more than 25,000 metric tons of oil escaped. The largest spill resulted from the collision of the tankers Atlantic Express and Aegean Star off Trinidad and Tobago on July 19, 1979; 300,000 metric tons of oil were spilled. Perhaps the best publicized incident of this sort was the grounding of the tanker Amoco Cadiz off Portsall, France, on March 16, 1978, in which 223,000 tons of oil escaped, much of it washing ashore and polluting beaches and fishing grounds off the coast of Brittany. The most common cause of major spills has been grounding, followed by collisions and hull failures. Although there obviously is some cause for concern of water pollution from pipelines, it would appear that marine transport of crude oil is in far greater need of attention.

5.3 REFINING

Refining has been restricted in this report to those processes of upgrading energy raw materials that are not carried out at or near the site of extraction. Thus the term is restricted generally to oil refining, commonly done at marine or inland terminal locations, and to several stages of nuclear-fuel processing carried on at various locations before the fabricated fuel is delivered to a nuclear-electric plant. Various processes of upgrading oil shale, tar sands, and coal normally carried out near the mine site have been discussed under Extraction (Section 5.1). In the case of oil-shale and tar-sand resources, a semi-finished product is transported by pipeline to a conventional refinery for further processing. The coal conversion processes that result in a change of physical state are discussed under "Conversion" (Section 5.4).

Figure 5.2.3. Bow view of supertanker, Mobile Arctic (130,00 tons), towering over tug alongside. Such large ships, when loaded and under way, require several kilometers to effect a change of course. (Photo courtesy of American Petroleum Institute.)

5.3.1. Oil refining

Waste products from refineries consist of oils, organic and inorganic chemicals (particularly acids, alkalis, sulfides, and phenols), and suspended solids. The major sources of these wastes are equipment leaks and spills, releases during shutdown or start-up of equipment, condensate from steam strippers, wastewater from crude-oil desalting equipment and storage tanks, washing down of equipment, blowdown from cooling towers and boilers, and chemicals from water-treatment processes. Other sources of waste are storm water and sanitary sewage. Spent clays from clay-treating units and bottom sediments from separators and traps are examples of waste solids. Clays are usually disposed of by dumping, but the other solids generally are pumped to sludge ponds as slurries.

Waste-treatment processes differ from refinery to refinery and are tailored to the processes employed and the physical layout of the plant. Most modern refineries segregate the waste so that similar wastes are collected for treatment in a single facility and wastes that require special handling are treated at their sources.

The wastes generally are segregated into four streams: oil-free waters, oily waters, process waters, and sanitary sewage. The oil-free waters include storm drainage, cooling-system and boiler blowdown, and water-treatment-plant chemical wastes. These waters generally contain only common inorganic constituents and are acceptable for direct discharge to streams without treatment. Oily waters include storm drainage and some cooling-system blowdowns that contain significant amounts of oil. These waters normally are processed in a gravity-type separator to remove oils and in a special-purpose separator if further cleaning is needed before direct discharge.

Process waters, which include desalted water, tank drawoffs, stripper condensates, pump-gland cooling waters, barometric-condenser water, and treating-plant wash waters, are usually heavily polluted with organics including, most prominently, phenols, cresols, and fatty organic acids. They normally go through a gravity-type separation and then receive secondary treatment. The various treatment processes employed, not necessarily all in a single refinery, include ponding, sedimentation, coagulation, filtration, skimming, pH adjustment, and chlorination.

Some chemical wastes require individual treatment. Acid and alkali wastes are neutralized by mixing them; sulfides are neutralized and then stripped out in an absorption tower; phenols are treated by aerobic bacterial processes and in trickling filters, oxidation ponds, and cooling towers. At most refineries, sanitary waste is collected in a separate sewer system and is discharged to municipal systems or, if that is impracticable, in a standard system on-site.

Normally, water-quality criteria applied to refineries have limits on maximum concentrations for biochemical oxygen demand (BOD), chemical oxygen demand (COD), oil, and phenol. BOD is the amount of dissolved oxygen used by microorganisms in the aerobic oxidation of organic matter in a sample of waste water at 20°C. The BOD is time dependent because the carbonaceous oxygen demand decreases with time, since the rate of biological activity decreases as the available food supply decreases. The most frequently used period for BOD is 5 days, and the parameter is usually given in mg/l. For example, a strong municipal waste water may have a BOD of 250 mg/l, while a phenolic-process water might be at 20,000 mg/l. The chemical oxygen demand is the amount of oxygen required for chemical oxidation of the organic matter to carbon dioxide and water by strong oxidants under acid conditions. No uniform relation exists between COD and BOD, except that COD must always be greater than BOD to the extent that nonbiodegradable organic matter is present. Process condensates may have COD-to-BOD ratios as high as 1.5 to 2. A fairly representative standard for refinery discharge would be BOD, 57 mg/l; COD, 286 mg/l; oil, 42 gm/l; and phenol, 1 mg/l.

It can be seen from the preceding discussion that emphasis is put on restricting releases of organic matter to surface waters. The inorganic constituents, mostly contributed by evaporative-cooling towers, cannot readily be removed and are accepted as part of stream loading. Because a great amount of inferior-quality water is handled in oil refineries, leaks from the surface and from holding ponds are not unheard of, and many older refineries especially are underlain by ground water contaminated from refinery operations.

5.3.2. Nuclear-fuel cycle

In most countries where fuel is produced for nuclear electric-power plants, the processing is carried out by the national government. Because the various processes have much in common with, and overlap production of, military raw materials, many of the activities are classified and little public information is available on specific details, including waste disposal.

As the term has been used earlier in this report, the "nuclear-fuel cycle" consists of three principal steps: (1) conversion of mill concentrates to uranium hexafluoride (UF_6), (2) enrichment of the uranium to a level of about 4 per cent U-235 in a gaseous-diffusion enrichment plant, and (3) conversion of enriched UF_6 to uranium oxide and its fabrication into fuel elements.

In step 1, uranium ore in the form of U_3O_8 received from mills is converted to UF_6 for input to the gaseous-diffusion enrichment process. This is basically a purification process in which impurities in the ore are removed. The principal liquid-waste-disposal practices affecting water qualtiy are releases of common ions from cooling-tower blowdown, fluoride and nitrate from chemical processes, and discharge to streams of low levels of uranium, radium, and thorium. The U.S. Department of Energy (1980) estimates these releases as follows, all normalized to an arbitrary energy production level of 29.3×10^6 kWh (10^{12} Btu) per year.

Releases to Water Bodies, Uranium Hexafluoride Conversion

Ion	Tons/yr	Element	Curies/yr
Fluoride	1.20	Uranium	1.8×10^{-6}
Sulfate	0.21	Radium 226	1.4×10^{-4}
Nitrate	0.01	Thorium 230	6.3×10^{-5}
Chloride	0.01		
Sodium	0.16		
Ammonium	0.07		
Iron	0.002		

In step 2, UF_6 is enriched to a level of 4 per cent U-235 by passing gaseous UF_6 through a series of porous barriers. Very high electrical input is required and there are large cooling requirements in connection with temperature control of the gaseous-diffusion process. Total circulating water requirements for a typical plant are 41×10^6 m^3 per day, with blowdown of 3,800 m^3 per day. Because of the large requirements, special treatment is used to reduce objectionable constituents from the blowdown, including chromium, zinc, and phosphate. According to the U.S. Department of Energy (1980), the releases to water bodies, calculated at an arbitrary power production level of 29.3×10^6 kWh per year, are:

Releases to water bodies, thermal diffusion enrichment

Ion	Tons/yr	Element	Curies/yr
Calcium	0.3	Uranium	8.3×10^{-4}
Sodium	0.4	Plutonium 239	5.1×10^{-12}
Sulfate	0.3	Neptunium 237	2.5×10^{-9}
Chloride	0.4	Ruthenium 106	5.9×10^{-5}
Nitrate	0.12	Zirconium-Niobium 95	1.3×10^{-5}
Iron	0.02	Cesium 137	9.5×10^{-7}
		Cerium 144	9.5×10^{-7}
		Technicium 99	4.4×10^{-3}

Step 3 consists of conversion of UF_6 to UO_2 and mechanical processing, including pellet production and fuel-element fabrication. The principal chemical steps include conversion of UF_6 to ammonium diuranate and its subsequent reduction to UO_2. Release of effluents to water bodies, calculated at an arbitrary level of 29.3×10^6 kWh per year, as reported by the U.S. Department of Energy (1980) are:

Releases to water bodies, fuel fabrication

Ion	Tons/yr	Element	Curies/yr
NH_3-Nitrogen	0.46	Uranium	8.3×10^{-4}
NO_3-Nitrogen	1.10	Thorium 234	4.2×10^{-4}
Fluoride	0.19		

The levels of direct discharge of inorganic constituents to streams are all within acceptable limits; the releases, of course, all add to stream loadings. Because dangerous radioactive chemicals are handled at all these facilities, special efforts are made to isolate all fluid waste streams, with special attention to avoiding any infiltration of fluids to the water table.

5.4 CONVERSION

"Conversion processes" in this report are those processes that change one form of energy into a radically different form. They include the conversion of the chemical energy of fossil fuels to electricity, the thermal energy of nuclear fission to electricity, natural geothermal energy to electricity, and the potential energy of streams to electricity and various processes for converting coal to more convenient, cleaner burning forms of fuel. The analogous processes of oil-shale and tar-sand upgrading are treated under "Refining" (Section 5.3).

5.4.1. Steam-electric power generation

The several methods of steam-electric generation are quite similar in their effects on water quality, the main difference being in the cooling process used, which is not determined by the form of energy input. Where once-through cooling is used, the entire thermal load of the power plant is transferred to the receiving water body. Where closed-cycle cooling, such as ponds or evaporative cooling towers, are used, the thermal load is transferred to the atmosphere on-site and certain chemicals are discharged. The chemical discharges differ to some degree among the processes. Coal-fired power plants have substantial ash-handling requirements and discharge degraded water used for that purpose as well as cooling-system wastes. Nuclear plants, in addition to cooling-system wastes, normally discharge low levels of radioactive elements. Geothermal plants, unless they can use the process of reinjection, must discharge large thermal loads together with trace metals and other constituents of the geothermal fluids.

5.4.1.1. Once-through cooling. The large amounts of water used in once-through cooling are generally treated only to the extent needed to control the growth of algae on condenser surfaces. Chlorine is commonly used for this purpose; however, chlorine may react with organics in the water stream to form chlorinated hydrocarbons, which at high levels are a hazard to human health. In addition, chlorine may react with ammonia found in many receiving waters to form chloramines, which are believed harmful to fish and other aquatic life. In plants where chlorine is used as an algicide, the residual levels in the cooling-water discharge generally are on the order of 1 mg/l (U.S. Department of Energy, 1981a, p. 41).

Once-through cooling systems are designed to release the thermal load to the receiving water body; the effect, of course, depends on the dimensions and state of flow in the water body. A large river or reach of shoreline with strong currents may show little effect. The main concerns center on streams where a substantial rise in temperature takes place. Most power plants raise the water temperature 8-11°C in the once-through mode; accordingly, if a stream is small or becomes quite warm in the summer season, a temperature safe for aquatic animals can easily be exceeded. Furthermore, organisms may become acclimated to a higher temperature regime, which raises both the upper and lower limits for the particular organisms. Then, if a plant is shut down in cold weather, the acclimated fish may die in large numbers. Other adverse effects include changes in dissolved oxygen and other dissolved gases, elimination of food sources, and decreased ability of fish to catch normal prey.

5.4.1.2. Closed-cycle cooling systems. Whether ponds and/or sprayers or cooling towers are used, all closed-cycle systems must release a certain proportion of the circulating water, termed "blowdown," to prevent the buildup of the dissolved-solids content in the circulating stream to a point where precipitation could occur on condenser surfaces, thus defeating the purpose of heat transfer. The ratio of blowdown to circulating stream is determined mainly by the quality of the inflow. Where the inflow, or makeup, is low in dissolved solids, proportionally less must be released than where the inflow is highly mineralized.

Closed-cycle systems must also face the problem of biological fouling, and chlorination is commonly used to control algae. As in once-through systems, the chlorine can combine with hydrocarbons in the water with similar adverse effects on the receiving stream of the blowdown. Scales of chemical precipitates are prevented by addition of inorganic phosphates, polyelectrolyte antiprecipitants, and organic-polymer dispersants. The first two are usually found in blowdown streams at concentrations of 2-5 mg/l, and the latter at 20-50 mg/l (US Department of Energy, 1981a, p. 41).

Coal-fired power plants, in addition, may discharge ash-handling waste waters, regardless of the cooling system employed. Most of the mineral matter in the coal is retained in the ash, and water used to slurry ash to disposal sites generally contains high levels of common ions as well as trace metals characteristic of the coal burned. It should be noted that both cooling-system blowdown and ash-handling wastes can easily be disposed of in arid climates by permitting the water to evaporate in lined ponds, thus avoiding the problem of contamination of streams or ground waters. Even in areas of higher rainfall, waste waters can be evaporated to dryness by sacrificing part of the energy output of the plant. Where such zero-discharge systems are used, the solid waste is periodically transferred to safe-disposal sites.

Most nuclear power plants are cooled by closed-cycle systems because of the large thermal loads to be dissipated. Furthermore, closed-cycle systems offer better means of control in the event of leaks of radioactive materials. Nuclear power plants release relatively little radioactive material in their liquid discharges, because the primary coolant that enters the reactor is kept isolated from the cooling-system supply. Despite this, mineral matter in the secondary stream is irradiated through close proximity to the more radioactive primary coolant. In addition, tritium generated in the fission process is released at low levels in the cooling-system blowdown. The annual releases of radioactive materials in the effluent streams of light-water reactors of 1,000 MWe capacity are reported by the U.S. Department of Energy (1981a, p. 232) as follows:

	Tritium (in curies)	Activation and fission products (in millicuries)
Pressurized-water reactors	300	120
Boiling-water reactors	25	15

Nuclear reactors produce high- and low-level radioactive wastes, which are solidified and packaged at the power plant for transport to safe disposal sites. The primary concern in safe disposal of both high- and low-level wastes is that the radioactivity should not enter circulating ground or surface waters and thereby pose a menace to human health. However, a meaningful discussion of radioactive-waste disposal is deemed to be beyond the scope of this report.

Geothermal fluids generally contain high levels of common dissolved minerals, as well as trace metals, including arsenic, mercury, selenium, and boron, at concentrations generally above those considered safe for direct discharge to ambient waters. Consequently, most newer geothermal plants rely on closed-cycle cooling, with disposal of waste waters, including cooling-system blowdown, by injection into the producing zone. Because extraction, electrical generation, and waste disposal are essentially different phases of a single process, they have been treated together in Section 2.1.5.

5.4.2. Hydroelectric power

Although hydroelectric power stations use the water passing through turbines only momentarily, the construction and operation of reservoirs may have important water-quality effects. Typically, water from storage dams is released through the power plant from considerable depth. In temperate climates this commonly leads to a complete change in the thermal regimes of the reaches immediately downstream from such dams, with cool-water fauna and flora replacing a warmer water biota. Furthermore, the design and operation of a reservoir may have significant effects on aquatic life through seasonal changes in turbidity, dissolved oxygen and nitrogen, dissolved-mineral content, and altered nutrient and organic matter regimes. These changes and their effects on downstream ecosystems depend to a great extent on thermal and chemical stratification in the reservoir and its relation to the depth at which water is withdrawn.

In reservoirs with constant throughflow or periodic turnovers of the water, the effects of stratification may be negligible, but in thermal regimes characterized by strong summer temperature stratification, the deep waters may become greatly depleted in oxygen, and this alone can have a severe effect on downstream aquatic life. In extreme cases, the deep waters may be completely depleted of dissolved oxygen, whereupon anaerobic bacterial processes predominate. This can result in marked lowering of pH, generation of methane, ammonia, and hydrogen sulfide, and can have drastic effects on the solubility of organic and inorganic constituents of the bottom sediments. The release of such waters, of course, can have a severe effect on biota downstream.

Another environmental effect of hydroelectric power facilities, generally beneficial, is the regulation of stream flow. Both maximum and minimum flows, to some extent, are moderated by hydroelectric dams. The effect is greatest at dams that have considerable storage capacity and is rather limited at run-of-river facilities.

Pumped-storage hydroelectric facilities are coming into wide use in hydroelectric systems as ways of storing surplus energy during off-peak times for use during peak times. In such facilities, water is lifted by a reversible pump-generator from a low-level reservoir to a higher level reservoir during off-peak periods and then is dropped through the turbine to generate electricity at peak demand periods. Such systems normally are used over a diurnal cycle; however, the same principle could be applied over a longer cycle, if the resource and economic situation warranted. As with most other hydroelectric facilities, the environmental impact is benign, and such systems have little effect on water consumption.

5.4.3. Coal-conversion processes

Coal-conversion processes have much in common with oil refining, oil-shale processing, and tar-sand processing. In each case, a solid or fluid raw material is subjected to a thermal-cracking process to drive off the volatile constituents, followed by one or more catalytic reactions aimed generally at adding hydrogen to the hydrocarbon molecules so as to produce a fuel of lower specific gravity and higher heat value than the raw material. Water is used for cooling (generally in several stages where condensation of gases is required), as a source of hydrogen for the chemical reactions, for gas cleaning, for boiler feed, for cleaning equipment, and for sanitary service.

The effects on water quality include the addition of conditioning chemicals and the concentration of dissolved load (resulting from cooling processes), the addition of dissolved mineral matter (resulting from the stripping processes), and the addition of large amounts of organic matter (resulting from the process steps in which water or steam comes in contact with the raw material or various hydrocarbon outputs). Coal-conversion plants employ high technology, and generally there are strong economic incentives to minimize energy losses of all types. Consequently, heavy emphasis is placed on recycling and reuse of water and on minimizing heat and chemical-waste discharges.

In coal-conversion plants, there are up to four key water effluent streams: a foul-process condensate recovered from the gasification step, intermediate streams recovered after gas purification or methanation, a cooling-system blowdown stream, and storm drainage and sanitary sewage. Organic contaminants, including phenols, cresols, and carboxylic acids, are present in large amounts in most processes, and most yield large quantities of ammonia derived from the nitrogen content of coal. Carbon dioxide is nearly always present and hydrolyzes to carbonic acid in the effluent. Hydrogen sulfide is usually present, in varying amounts depending on its concentration in the input coal.

In water treatment, much of the organic matter may be removed by biological processes. When decomposition takes place under aerobic conditions (that is, in the presence of free oxygen), the products are carbon dioxide, water, and bacterial sludge. Thus, the dissolved organic matter is partly oxidized and partly converted into hydrocarbon cell material. When oxygen is not available, anaerobic bacteria use combined oxygen of the organic and inorganic matter in solution for their life cycle and produce carbon dioxide, methane, hydrogen sulfide, and inert products. Many oranic compounds in coal-conversion process waste streams cannot be (or can only very slowly be) decomposed by bacteria; most, however, can be chemically oxidized to carbon dioxide and water. Accordingly, most water-treatment facilities rely on a combination of bacterial and chemical processes to break down dissolved organic matter; hazardous inorganic constituents generally are removed by chemical precipitation reactions.

Many inorganic solutes, including the anions, chloride and sulfate, and sodium, are not readily removable at reasonable cost and therefore are discharged to receiving waters or to zero-discharge systems. This is generally the case with concentrations of dissolved solids through cooling-system evaporation, for example.

There are literally dozens of coal-conversion candidate processes, all differing in various aspects of chemical and physical methodology. To treat each technology in depth is beyond the scope of this report; however, the interested reader is referred to Probstein and Gold (1978), who give in-depth descriptions of individual methodologies.

5.5 SEVERITY OF EFFECTS AND EFFECTIVENESS OF CONTROLS

Table 5.5 summarizes for all the energy processes described in this report the likelihood (frequency), areal scale, time frame for correction, severity of impacts on water quality, and the effectiveness of control measures normally applied to protect streams and ground waters. The outstanding problem areas identified are underground coal mining, hot-water geothermal systems, uranium mill-tailings disposal, marine oil-tanker transport, and low-level radioactive-waste disposal.

The column headed "Frequency and areal scale" indicates whether the scale of water-quality impact generally is of regional scope or local scope by the placement of the letter indicating frequency (or likelihood of occurrence) as high, medium, or low. Thus, for each line entry a letter appears in either the regional or local column, but never in both. The column headed "Time frame" indicates by letter the general period of duration of problems identified. In similar fashion, the column headed "Severity" indicates by a numerical rating the general level of maximum adverse impact; however, in many specific locales the impact may not reach this upper limit. Finally, the column headed "Effectiveness of controls" indicates the state of the art with respect to control measures in general application, which may not necessarily represent the "best available technology." For example, in the case of the impact of underground coal mining on surface waters, the control measures are shown as poor, because the most widely applied controls are limitations on the pH and iron content of mine point discharges to surface streams. By adjusting the pH through lime treatment, the mine operator may be able to comply with pollution-control regulations without addressing the broader problem of stream degradation due to increased dissolved-mineral content, which is in no way reduced by lime treatment. Moreoever, as in the Appalchian coal fields of the United States, enforcement of discharge limitations on active mining operations fails to address the lingering effects of worked-out and abandoned mines, which continue to contribute degraded waters to streams long after their closure.

Coal mining is a widely dispersed activity in many countries, underground mining is not subject to public observation, and adverse impacts on water resources are slow to take effect. The effects on water quality in the form of inorganic degradation of surface and ground waters and disruption of aquifers are widespread and long-lived. Problems are compounded because of the wide availability of coal and the relative simplicity of mining and shipping. Competition for markets is sharp, and the supplier who can minimize cost of production is likely to get business. In view of this economic climate, enforcement of enlightened water-pollution practices is most difficult. Moreover, the predominant forms of underground coal mining, room-and-pillar and complete recovery, make water pollution almost inevitable. There is great opportunity for improved control technology to reduce the adverse chemical impacts of underground coal mining on water resources, but in most countries little headway is being made. With increased coal production foreseen over the next two decades, impacts on the quality of streams and ground waters are more likely to increase than to be remedied during the period.

In hot-water geothermal systems, the major problems involve safe handling and disposal of the high dissolved-mineral content of the geothermal waters, particularly the toxic trace metals. Hot saline waters are extremely corrosive; therefore leaks are frequent in even the best managed operations. The favored solution to disposal of geothermal brines is generally reinjection to the producing zone. However, this is made difficult by the tendency for the dissolved minerals to precipitate when the heat is extracted from the fluid. Commonly, precipitation ponds and cation-exchange processes are used to improve the effectiveness of injection, but this requires extensive holding ponds and increases the potential for escape of fluid from the ponds. The most severe effect is likely to be degradation of local ground waters, which subsequently may discharge to the surface. Chemical processes are available to control completely contamination by geothermal waters, but usually their cost makes geothermal conversion unattractive. This may be the main factor restraining severe contamination. When developers must correct the adverse aspects of their efforts, they are likely, in this case, to forego development. In fact, the areas of likely ground water contamination have, for the most part, already been contaminated by natural geothermal activity; it is the off-site effects that are likely to be the most damaging.

The atomic age burst on the world quickly only 40 years ago, and many adverse environmental impacts of nuclear-energy development simply went unnoticed in society's haste to reap the benefits of nuclear energy. Safe disposal of mill tailings was one aspect of nuclear hazards that was overshadowed by larger concerns for many years and has only recently been identified as needing greater attention. Technologies are available to alleviate the adverse impacts on water resources, but all involve considerable additional cost in handling of new materials and investments in correcting previous deficiencies. The key to safe disposal is relying on dry storage rather than the presently used wet disposal and minimizing access of water to tailings piles by the use of impermeable cappings and linings. The use of dry storage, of course, implies the separation of water and tailings in mill effluents, treatment of the fluid to make it suitable for direct discharge to ambient waters or zero-discharge designs to dispose of water by evaporation, and safe ground disposal of the solid residue from evaporation ponds.

The general rise in costs of alternative fuels since 1973 should have a salutory effect on improving methods of uranium-mill waste discharge, in providing funds for improved disposal practice without making nuclear fuel inordinately expensive. Thus, the outlook for improvement is encouraging over the rest of this century.

Most marine oil-tanker spills have resulted from collisions, groundings, and hull failures. The latter factor suggests a need for better control over safety through stiffer design standards and follow-up safety inspections. Collisions and groundings suggest a need for better navigation and warning devices and better seamanship. Here, too, the rise in oil prices should have a salutory effect. In the collision of the Atlantic Empress and the Aegean Star in July 1979, when 300,000 tons of oil escaped, the loss of oil alone at 1983 prices

Table 5.5. Water-quality impacts, by energy processes

Note: Frequency: H = high, M = medium, L = low. Time frame: L = longer than 10 years, M = 1 to 10 years, S = less than 1 year. Severity: 5 = direct threat to human life, 4 = hazardous to human health, 3 = severe economic damage, 2 = damage to biota, 1 = aesthetic or other intangible harm.

Process	Water-quality impacts	Frequency and areal scale		Time frame	Severity	Effectiveness of controls
		Regional	Local			
Extraction and on-site processing						
Coal mining						
Underground mining	Most damaging problems are acid mine drainage and disruption of aquifers, which affect pH, dissolved solids and specific ion content, and thus impair utility of streams and ground waters for other uses.					
	a. Surface waters	H		L	3	Poor
	b. Ground waters		M	L	3	Ineffective
Surface mining	Surface disturbance results in high sediment transport potential. Discharge from mines may impair water quality through increase in dissolved solids and specific ions.					
	a. Surface waters	H		M	3	Fair
	b. Ground waters		L	S	1	Good
Beneficiation	Release of chemical and physical treatment materials to streams can impair water quality. Leaching of solid wastes results in pollution similar to acid drainage.					
	a. Surface waters	H		M	2	Fair
	b. Ground waters		L	M	1	Fair
Oil and gas extraction						
Primary recovery	Principal problems are handling of saline waste waters. Leaks in casings, pipes, and storage ponds can release brines to ground waters and streams.					
	a. Surface waters	M		S	2	Good
	b. Ground waters		M	M	3	Fair
Secondary and tertiary recovery	Principal concerns are escape of oil and formation waters through casing, pipe, and storage tank leaks releasing organic and inorganic contaminants to the environment.					
	a. Surface waters		L	S	2	Good
	b. Ground waters		L	L	3	Good
Offshore operation	Blowouts with resulting massive oil contamination are a rare but catastrophic problem.	L		S	3	Good

Table 5.5. Water-quality impacts, by energy processes--Continued

Oil-shale extraction and processing	Most significant concerns relate to potential for escape of noxious organic and inorganic contaminants to streams. Disruption of aquifers likely, low hazard due to limited occurrence of oil shale.					
Underground mining	Concerns center on disruption of aquifers and disposal of sometimes saline dewatering by injection.					
	a. Surface waters		L	L	1	Good
	b. Ground waters		L	L	1	Good
Surface mining	Surface disturbance results in high sediment transport potential.					
	a. Surface waters		L	S	2	Excellent
	b. Ground waters		L	M	1	Excellent
Surface retorting	Concerns center on potential for escape of organic and inorganic contaminants from plant site due to accidental leaks and spills. A more significant concern is escape of contaminants from waste piles through leaching.					
	a. Surface waters		M	L	2	Good to Fair
	b. Ground waters		L	L	1	Good
In-situ recovery	Underground effects mainly involve contamination of ground waters by organic and inorganic compounds produced in combustion; where applicable surface effects are similar to those of surface retorting.					
	a. Surface waters		L	S	2	Excellent
	b. Ground waters		M	L	3	Untested
Tar-sands extraction and processing						
Surface mining and processing	Main conerns are accidental release of organic contaminants to streams and potential for failure of waste impoundment structures leading to massive downstream damage from fine waste.					
	a. Surface waters		M	S	2	Good to Fair
	b. Ground waters		L	L	1	Untested
In-situ recovery	Concerns center on potential for escape of noxious organic and inorganic chemicals to ground waters.					
	a. Surface waters		L	S	2	Untested
	b. Ground waters		M	L	3	Untested

Table 5.5. Water-quality impacts, by energy processes--Continued

Geothermal extraction						
Vapor-dominated systems	Main concerns are escape of noxious inorganic contaminants to ground water from waste disposal, blowouts, and leaks in casings and pipes.					
	a. Surface waters		L	S	2	Good
	b. Ground waters		M	L	1	Fair
Water-dominated systems	Main problems involve escape of noxious and toxic constituents of thermal waters to surface and ground waters from production operations, waste disposal, blowouts, and leaks in casings and pipes.					
	a. Surface waters	H		M	4	Fair
	b. Ground waters		H	L	3	Fair
Uranium mining and milling						
Underground mining	Escape of radioactive and other inorganic contaminants to the environment through disposal of dewatering waste and escape from tailings ponds can seriously impair downstream water uses.					
	a. Surface waters	M		S	4	Good
	b. Ground waters		M	L	4	Poor
Surface Mining	Some concern about high sediment transport potential, but main source of concern is potential for radioactive contamination of streams and ground waters through leakage from tailings disposal ponds.					
	a. Surface waters	M		S	4	Good
	b. Ground waters		M	L	4	Poor
Solution mining	Main concern centers on escape of radioactive and inorganic process chemicals to off-site ground waters.					
	a. Surface waters		L	S	1	Excellent
	b. Ground waters		L	L	4	Excellent
Transportation						
Coal slurry lines	Main concern centers on pipeline breaks and the potential for contamination of streams.					
	a. Surface waters		L	S	2	Excellent
	b. Ground waters		L	L	1	Excellent
Oil pipelines	Most significant problems are pipeline breaks and resulting oil pollution of streams.					
	a. Surface waters		M	S	2	Fair
	b. Ground waters		L	L	1	Good
Oil tankers	Escape of oil to marine environment as result of shipwrecks can be catastropic to marine life over wide areas.	L		M	3	Poor

Table 5.5. Water-quality impacts, by energy processes--Continued

<u>Refining</u>						
Oil refining	Controlled release of waste water and accidental releases of organic and inorganic contaminants are most significant issues; concerns center on impairment of water supplies of other water users.					
	a. Surface waters		M	M	3	Good
	b. Ground waters		L	L	3	Good
Nuclear fuel cycle	Accidental releases of radioactive materials to surface and ground waters from processing and reprocessing plants are main concern; both high- and low-level waste disposal also have potential for escape of radioactivity to the water environment. Controlled release of non-radioactive inorganic chemicals adds to chemical load of receiving waters.					
	a. Surface waters		M	S	4	Good
	b. Ground waters		M	L	4	Fair to poor
<u>Conversion</u>						
Fossil-fueled steam electric generation	Controlled release of cooling-system blowdown to streams and/or leakage from cooling ponds adds dissolved solids and treatment chemicals to stream loads. Once-through cooling contributes to thermal pollution.					
	a. Surface waters	M		M	2	Good
	b. Ground waters		L	L	1	Good
Nuclear steam-electric generation	Small controlled releases of radioactive materials and discharge of cooling-system blowdown adds radioactivity, dissoloved solids, and treatment chemicals to stream loads. Accidental release of radioactivity through reactor containment failure could endanger human life over wide area.					
	a. Surface waters	M		M	5	Good
	b. Ground waters		L	L	4	Good
Geothermal electric generation	Disposal of waste and condensate containing noxious inorganic compounds and thermal load to streams impairs downstream uses and damages aquatic life.					
	a. Surface waters	H		M	4	Fair
	b. Ground waters		H	L	3	Fair
Hydroelectric generation	Changes in stream temperature and dissolved gases due to storage and reservoir releases seriously alter the aquatic environment.					
	a. Surface waters	H		L	2	Fair
	b. Ground waters		L	S	1	Good

Table 5.5. Water-quality impacts, by energy processes--Continued

Coal Conversion Processes	Controlled release of cooling system blowdown and accidential releases of organic and inorganic contaminants, as in oil refining and with similar concern about impairment of other water uses.					
	a. Surface waters		M	M	3	Good
	b. Ground waters		L	L	3	Good

would have been US $65 million over and above the loss of ships and life. It would appear that there is ample economic incentive to avoid such spills, but regulation is difficult because of common ownership of the high seas by all nations. Disposal of low-level nuclear waste, like disposal of uranium-mill tailings, is one aspect that received little attention in the early atomic age. It was recognized that disposal of toxic, highly radioactive waste from nuclear power plants would require great care, but the less concentrated wastes that are generated in other nuclear processes were (in the United States at least) all swept into the general category of low-level wastes. In this category were included small concentrations of toxic, long-lived wastes. The title "low-level waste" sounded innocuous, and methods of disposal similar to those used at sanitary landfills were permitted for a wide variety of nuclear trash. Disposal sites were selected in impermeable materials, and boxed waste was placed in trenches and subsequently covered over with earth from the trenches. However, questions as to the effectiveness of containment have arisen at several disposal sites. Of the six sites in the United States that were licensed for commercial use (that is, non-governmental, low-level waste disposal), only two are currently (1983) allowed to operate. Responsible technical opinion is focusing on alternative methods of waste disposal, mainly involving greater initial processing to separate radioactive materials, especially the long-lived isotopes, from the general mass of trash and to solidify such materials for long-term storage. No doubt there will also be improved government surveillance over waste-storage operations in the future to assure effective long-term containment.

The outlook for the correction of problems in low-level radioactive waste disposal is encouraging, and the most severe problems should be remedied within a few years. As even low-level wastes occupy relatively little volume and are easily detectable, they are susceptible to proper control to prevent their escape to streams and ground waters.

Other energy processes present fewer serious problems with respect to their effects on water resources. Many problems have already been corrected or at least are amenable to straightforward, economically feasible control. In many countries, release of noxious and hazardous materials to streams has been greatly reduced over the past 20 years and further restrictions on waste releases will probably be put into place over the next two decades. Controls on sediment transport, for example, are simple and highly effective, and enforcement surveillance requires only regular monitoring. High technology processes usually involve such large capital investments that control of effluents becomes a relatively small cost element. Moreover, in most such processes there are strong economic incentives to avoid gross contamination of water.

One area, however, that offers little hope for improvement over the balance of this century is in the concentration of mineral matter through evaporative cooling. With environmental concerns forcing once-through cooling systems out, closed-cycle cooling will soon predominate in disposal of thermal waste in all energy processes. Because the concentration by evaporation in closed-cycle systems is about twice that of once-through systems of comparable capacity, the concentration effect will become more of a problem. There is no methodology for removing mineral loads from large cooling streams that is even remotely within economic reach; hence, there will be no relief short of zero discharge systems, which are impracticable in most cases. In arid climates, where zero discharge systems are practicable, the higher evaporation they entail is objectionable because it wastes valuable water resources.

6 New methods of water use in the energy sector

The principal new uses of water in the energy sector that can be predicted with confidence include hydraulic coal mining and integrated in-mine transport systems to move the coal from the working face to the surface, coal-slurry pipelines integrated with marine slurry transport and delivery from ship to customer by surry pipeline, and the so-called hydrogen-fueled transport systems. The latter, however, probably will have to await the emergence of much cheaper sources of energy, perhaps until fusion power becomes a practical reality.

Slurrying coal from producing areas to demand centers appears to be on the threshold of great expansion as the most efficient way of moving large tonnages of coal over long distances. The current United States shipments by this method are about 5 million metric tons per year via the Black Mesa slurry line (Figure 2.2.1). This will see a five-fold increase when the ETSI slurry line from Wyoming to Louisiana goes into operation (scheduled for 1985). Other slurry lines currently planned in the United States would have combined capacities of nearly 80 million metric tons. As currently operated, this would require roughly 80 x 10^6 m^3 of water. Although most of this water is separated from the coal at the terminus of the pipeline and can be reused or discharged, this amount of water represents an element of consumption from the area of origin of the water.

Other slurry techniques that are under serious consideration include integrated marine transport by coal-slurry tankers and slurry-line delivery systems in coal-importing countries. Such systems offer many advantages over conventional transport arrangements. Ships equipped for slurry handling require only pipeline connection to a simple mooring facility for loading and discharge. The need for docks, cranes, ship loaders and unloaders (as well as conveyers, stackers, and reclaimers) is eliminated. Drawing on technology developed in handling offshore oil shipment, slurry pipelines could use offshore terminals for loading the slurry aboard tankers and unloading it at destinations. A slurry export or import terminal would have two or more pipelines running to a shore-based storage area for loading or unloading the ship's cargos. One line would handle the slurry and the other would return the carrier fluid. This sort of terminal represents an alternative to development of deep-water ports with conventional loading systems.

Speaking at the Fifth International Technical Conference on Slurry Transportation, Las Vegas, USA, May, 1980, D. Hoogendoorn of the Netherlands reported on a project for handling coal-slurry imports at Rotterdam, Netherlands. He estimated a savings of 25 per cent over the cost of a conventional transportation system.

In regard to water use, current slurry-line designs require equal amounts of water and coal, and there is no reason to expect this ratio to change. Thus, export of coal will entail export of equal volumes of water in international trade, as there would be little economic incentive to return the carrier water to the country of origin.

Hydraulic coal mining and slurrying machinery are available from a number of manufacturers and are in active use on a commercial scale in several mines. One example is the underground hydraulic mine of B C Coal, Ltd. at Sparwood, British Columbia, Canada (Schneiderman, 1981). Here a 15 m thick seam of bituminous coal that dips 35° to 50° is being mined using hydraulic excavation and transportation to the surface. A hydraulic monitor supplied from surface pumps shoots a stream of water at the coal face at pressures of 110 to 120 kg/cm^2 at a rate of 7,600 l/min. Coal is broken away by the force of the water jet and is drawn into a feeder breaker, which breaks it into pieces less than 150 mm for transport. Steel flumes move the coal in a water stream by gravity from the coal face to an underground dewatering station where plus and minus 9 mm fractions are separated. The fine fraction is slurried to the surface in a hydraulic system, where it is dewatered for conventional transportation; the coarse fraction is raised to the surface by conventional systems.

The hydraulic miner cuts a panel 15 m high by 26 m wide in a retreat mode; after each 12 m of retreat, the monitor is moved 12 m back toward the mouth of the sublevel for another cut. Average production by the hydraulic miner has been 1,741 tons per shift and has totaled 430,000 tons during 10 months of operation in 1980. While operating, the miner is able to produce up to 6 tons per minute. The mining and transport system is based on recirculation of water, but there is a net makeup requirement of about 20 per cent. Evidently considerable water leaves the mine with the coal; the difference apparently is lost by percolation into what is described as fractured and faulted strata.

In energy conversion processes, the likely future developments all point toward lowered water use and consumption. Much research and development effort is going into improving overall thermal efficiency above the present 35-40 per cent upper limit for steam-electric generation. One approach that appears to have considerable merit is the use of a combined cycle in which gas from a natural-gas line or from an on-site coal gasification plant is used to drive a gas turbine, and the hot exhaust gases from the gas turbine are used to generate steam for a conventional boiler-steam turbine generator system. It should be possible to achieve overall thermal efficiencies close to 50 per cent in this way because the waste heat of the gas turbine is put to further use. Because the gas turbine has no cooling water demand, the consumption of water would be reduced in proportion to the increase in thermal efficiency of the combined system.

Another electrical generating cycle that holds great promise for the future is the fuel cell, in which the chemical energy of a hydrocarbon fuel, such as gas or oil, is converted directly to electricity without combustion. Because the conversion takes place without combustion, the thermal efficiency is much higher. Moreover, there is no transfer of heat to water to drive a turbine. Consequently, the thermal efficiency of the system is higher and cooling requirements are lower than in the conventional boiler-steam turbine system. Systems that use photovoltaic cells to convert solar energy directly to electricity are far too expensive and inefficient for large-scale electric generation at present, but they have been used to meet small electrical demands at remote sites where conventional power systems are uneconomic. It is possible that production costs may be lowered to a level at which this mode of generating power will become more practical. Because there is no cooling demand in such a system, its broader application would reduce water use in electrical generation.

Still another technology that may have some potential in favorable sites is the use of an array of solar collectors which transfer the energy collected to water for input to a conventional steam-turbine cycle. This process presumably would have low thermal efficiency due to the low temperature of the input water, so the unit consumptive use of water probably would be relatively high, as is true of geothermal power. However, other design and operational problems raise even more serious questions as to the practicality of the systems. For example, it has been estimated that the metal collectors required to supply a 1,000 MW power plant would occupy more than 2 million hectares.

The use of hydrogen as fuel for internal combustion engines and for space heating has been much discussed by futurists, and hydrogen clearly would have many environmental advantages, the fact that it is clean burning being the most evident. However, for hydrogen fuel to become a practical reality, the cost of fossil fuels or electricity would have to be greatly reduced. Water is an obvious source of hydrogen, but to produce hydrogen from water by known processes requires extravagant energy input to break the oxygen-hydrogen bond. The most practical direct way of producing hydrogen from water at present is by electrolysis; however, operating a vehicle with this source of hydrogen would be far more costly than using hydrocarbon fuels. The lowest cost method for producing hydrogen currently is using petroleum as input to a chemical reaction that yields hydrogen. Most synthetic fuel conversion processes are moving in the opposite direction, that is, adding hydrogen to existing hydrocarbon molecules. Clearly that is the most cost-effective route to clean-burning fuels in the present economic climate. However, should fusion power generation reach practicality, the world may once again see a cheap source of electrical energy, which could make hydrogen directly from water for a great variety of uses.

7 Planning alternatives

Because water plays a vital role not only in energy production but also in the natural environment, planners must consider both the availability of water and the potential environmental effects on water resources when making energy development decisions. In the following paragraphs some of the main alternatives are discussed.

With respect to water availability, three general cases exist, with the related principal alternatives.

1. Water plentiful. The planner has a wide choice of location for water-using facilities. The decision depends largely on the economics of product transport (such as electrical transmission, product pipelines, rail, or water transportation). Mine-mouth production may be selected as the most cost-effective type of development, or transportation of energy raw material to the point of ultimate use may be preferred.

2. Water limited in quantity and/or costly to develop. More constraint is placed on the planner and the cost of supplying water to the facility assumes a larger role in the decision, but economics still dominate the decision-making process. The planner may opt to transport the energy raw material to the point of use for conversion, or may carry out conversion at an intermediate location where water is readily available at lower cost (Figure 7). As a general rule, mine-mouth development is precluded in such cases, except where great bulk of raw energy material makes transport impracticable.

3. Water is deficient or unavailable due to legal constraints. The range of options is expanded in an effort to produce the product with an economic gain. Some of the principal alternatives are:

   a. No development.
   b. Use of air cooling or partial air cooling at greater captital cost and loss of efficiency.
   c. Importation of water from other drainage basins.
   d. Purchase of local water from other users.
   e. Development of unappropriated local water resources by construction of reservoirs and conveyance works or by drilling of wells.
   f. Export of raw material to point of use.
   g. Export of water to intermediate site for processing where water is available.

In regard to water pollution impacts, two cases generally apply: where water pollution is a minor concern and where it is a major concern. The main areas of concern are disposal of thermal waste, disposal of chemically degraded water, disposal of solid noxious waste, disposal of hazardous to human health, and mobilization of sediment and soluble minerals through land disturbance. Each concern presents several options to planners.

1. Disposal of thermal waste.
   a. Site plant on large water body that can accept heat with minimal impact.
   b. Use closed wet-cooling system or combined wet/dry system.
   c. Use closed system in zero-discharge mode.
   d. Use all air cooling.

Figure 7. Mojave coal-fired power plant (1,500 MW) in southern Nevada, USA. Coal is supplied from Navajo mine, 440 km east, via coal-slurry pipeline. Note forced-draft cooling towers to left of plant, which account for water consumption of 22.4 million m^3 per year; remaining 5.6 million m^3 is consumed in blowdown evaporation from ponds, right of plant, and evaporation of a portion of coal-slurry water that is not economic to reclaim for cooling use, from pond in left foreground. Plant was sited here because of availability of cooling water from Colorado River, upper center. (Photo by U.S. Bureau of Reclamation.)

2. Disposal of chemically degraded water.
   a. Site plant on ocean, large lake, or major river with large capacity to absorb waste.
   b. Use zero-discharge system and transport solid residue to safe disposal site.
   c. Use in-plant purification system to upgrade waste to acceptable level for direct discharge.
   d. Use underground injection to zones where potention for contamination of useful water resources is negligible.

3. Disposal of solid noxious wastes.
   a. Dispose of waste locally, taking care to prevent later mobilization of soluble materials.
   b. Transport waste to safe disposal site.
   c. Dispose of waste underground in mined-out areas or other cavities.

4. Disposal of wastes hazardous to human health.
   a. Transport waste to safe disposal site with long-term care to ensure that hazardous components will not enter the hydrological cycle.
   b. Dispose of waste underground in mined-out areas or other cavities as solids.
   c. Dispose of waste as liquid in deep isolated permeable zones by injection, with monitoring to ensure that harmful components will not escape to damage usable water resources.

5. Mobilization of sediment and soluble mineral matter through land disturbance.
   a. Minimize area subject to erosion at any given time by proper planning of mining and reclamation work.
   b. Employ sediment control to keep sediment on the area of operations and provide for prompt revegetation of the disturbed area.
   c. Use water treatment as appropriate to reduce contamination from direct discharges.
   d. Limit leaching and underground drainage by adequate surface drainage, sealing of underground openings, and use of impermeable coverings.

The foregoing is not an exhaustive list of policy alternatives but is offered as a sampling of the major options generally available to planners.

8 Summary and conclusions

The first part of this report discusses the water-using elements of the energy sector and describes water use and water consumption for each element. In terms of gross water use, that is, the total amount of water that is passed through the facility, the largest use by far is by hydroelectric plants, which can pass the full flow of major rivers through their turbines. This, however, has little impact on the amount or quality of the resource, because the water is used only momentarily and the water is returned to the river with no significant change in quantity or quality.

A far more significant water use is that of steam-electric power plants, which, because of fundamental design limitations, must dissipate about two-thirds of the energy input of the system to the atmosphere as waste heat. Water is nearly always used as the coolant in such plants, and evaporation of part of the water is the chief means of transferring heat to the atmosphere. Some 78 per cent of the world's electric power is generated in steam-electric plants; thus a very large heat load must be dissipated in this way. The majority of steam-electric plants use once-through cooling systems in which the waste heat load of the power plant is transferred to a moving stream of water, which is returned to the source virtually unchanged except for the added heat. The heat is dissipated to the atmosphere in large part by increased evaporation from the water surface. In the industrial countries of Europe and North America, such use of water far exceeds other industrial uses of water. In the United States, for example, withdrawals for steam-electric power generation account for roughly half of the withdrawals for all uses, excluding hydroelectric power.

A growing proportion of the world's steam-power plants are employing closed-cycle cooling systems, in which the waste heat is dissipated to the atmosphere on-site by means of cooling towers, ponds, and sprayers. This method avoids the serious adverse effects on receiving waters of once-through cooling but results in significantly greater water consumption, because in closed-cycle cooling systems the heat rejected to the cooling system must be disposed of almost entirely by evaporation, while in once-through cooling a significant part of the heat is dissipated by convection and radiation to the atmosphere and by conduction to the bed of the water body. Throughout the world, water consumption by steam-electric power plants far exceeds consumption in other elements of the energy sector and comprises nearly three-fourths of the total water consumption by energy processes.

Next in importance in terms of water use and consumption is oil refining, which accounts for 21 per cent of world's water consumption in the energy sector. As in electric-power generation, the major water use is in evaporative cooling to dissipate waste heat, although small quantities of water are used as a source of hydrogen in a number of refining processes.

Water used and consumed in extraction of raw materials is modest compared to that in evaporative-cooling processes. The largest single element of water consumption is in oil production, where water and steam are injected into producing reservoirs to mobilize the oil and displace it toward production wells. A volume of water equivalent to the oil displaced remains in the reservoir when the operation is completed and thus represents a consumptive use of the water. In other extractive operations, water is used widely for dust control, for separation processes, and as a carrier in waste disposal; however, most water used in those ways returns to streams and ground waters.

Although electric-power plants and oil refineries together account for most of the world's water consumption in energy processes, this is offset to a considerable degree by the fact that much of the water consumed in those processes is from the sea or estuaries, and thus the energy-production users are not in competition with other water users for supplies. In the United States, for example, saline water constituted one-third of the total withdrawals of water by steam-electric plants in 1975. This reflects the fact that the ocean is an ample source of water for once-through cooling systems and that many of the large electrical load centers of the United States are in coastal cities. The oil refineries of the world, for the most part, are on the seacoasts or estuaries near marine tanker terminals; consequently, a large proportion of their water consumption is from such water bodies. Inland refineries are located mainly in the United States, Canada, and the USSR to serve inland demand from nearby producing fields.

Values of unit consumptive use of water were derived for each energy process, expressed as liters per ton oil equivalent (l/toe), or liters per kilowatt hour (l/kWh) for electrical output. The unit consumptive use by electric generation exceeded that of most other energy processes by an order of magnitude or more. For example, the average unit consumptive use of water in oil refining was estimated at 1,200 l/toe, compared with 35,000 l/toe (3.0 l/kWh) for nuclear steam-electric generation. The highest value was that of geothermal power at 174,000 l/toe (15 l/kWh), reflecting the low overall thermal efficiency of geothermal-electrical generation. The only large value outside of cooling processes was that of oil production, almost entirely for water flooding, of 1,100 l/toe.

Unit-consumptive-use values were applied to data on world energy production and consumption obtained from publications of the United Nations, the World Bank, and OECD to arrive at world water consumption by energy processes and consumption within economic groups--the OECD countries, the centrally planned economies, OECD-Europe, and the developing countries. Similar estimates were presented individually for the United States and Canada. Water consumption by energy processes throughout the world totalled 16 x 10^9 m^3 in 1980, of which 11.3 x 10^9 m^3 was accounted for by steam-electric power generation and 3.3 x 10^9 m^3 by oil refining.

Unit-consumptive-use values were modified to reflect expected technological changes in the next two decades, and these revised values were applied to recent energy projections for the year 2000. On this basis, it is estimated that world water consumption in the energy sector will increase to 38 x 10^9 m^3 by 2000. Major increases are expected in electrical generation (particularly by nuclear plants) and in water consumption in oil production, due mainly to increased use of secondary- and tertiary-recovery processes.

The effects of energy developments on water quality were described for each energy process and the information was compared in a table summarizing information on frequency and areal scale of effects, the time frame for alleviating adverse effects, the severity of impacts, and the effectiveness of controls. The leading problem areas identified were underground coal mining, uranium milling, geothermal development, and nuclear power plants in the event of a containment failure. Underground coal mining causes, and for the foreseeable future will continue to cause, gross inorganic contamination of streams in areas where acid mine waters discharge. The problem is widespread, dispersed, and not readily susceptible to technological solution.

Uranium milling results in great quantities of waste rock material (or "tailings") which can pose a threat of inorganic and radioactive contamination to streams and ground waters, particularly where the tailings are disposed of wet to ponds, which can leak contaminants to ground waters. Technological solutions are available at added cost and are being implemented in many areas. However, there are many large tailings dumps in sensitive locations which need prompt attention. Nuclear power plants are a source of concern because in the unlikely event of a failure of the reactor containment, the subsequent radioactive contamination would pose a serious hazard to human life and health over a broad area. In the nuclear fuel cycle, the danger of contamination of surface and ground waters from low-level waste disposal is of modest concern because of the past practice of disposing of some highly toxic, long-lived materials, albeit at low concentration, along with less hazardous trash in disposal areas similar to land-fills. Technological solutions are available for this problem at substantially higher cost. The disposal of high-level radioactive waste is not addressed in this report, because there are no plans to dispose of high-level wastes in ways that will have an impact on the hydrologic environment. Indeed, one of the primary criteria in disposal of high-level waste is to provide assurance that such radioactive materials will be isolated from the water environment for tens of thousands of years.

Geothermal development of water-dominated systems poses problems mainly because the hot waters are excellent solvents and dissolve a wide spectrum of contaminants in the earth. Some such waters are highly saline and that constitutes a disposal problem in itself. Of perhaps greater concern is the high content of toxic metals in most geothermal fluids. For most geothermal waters, the only technological solution for disposal is a zero-discharge system, either by reinjection of exhaust brines to the producing zone or by complete evaporation of fluids, with safe land disposal of the solid residues. In many settings these options may not be practicable, and this could forestall many promising developments.

Potential new methods of water use in the energy sector are mainly in the transportation field, where coal-slurry pipelines should have major impact in the short range. Coal-slurry lines are operating on a limited scale in the United States at present, and major expansion of this type of transportaion is foreseen for moving coal from producing areas to water terminals. The technique lends itself well to integrated land and marine transport systems in which a slurry pipeline discharges to a marine tanker, which transports the slurry to a receiving terminal, where the slurry is off loaded to another slurry pipeline for delivery to an inland consumer. Technically, the concept is feasible at present; however, wide application will depend on the economics of the integrated systems. Water is expected to find much additional use in coal mining, where hydraulic excavating by high-pressure jets holds considerable promise, together with integrated systems of slurrying the coal from the mining face to the surface. Such systems are being tested on a commercial scale at several locations.

Among the current large uses of water for electric-power generation and oil refining there may be significant reductions in unit consumptive use incidental to improvements in overall thermal efficiency in the interest of energy conservation. However, this will be offset over the short term by increased use of closed-cycle cooling systems.

Because the cost of water rarely determines the siting of an energy facility, planners generally consider cost and availability of water and waste disposal along with a host of other considerations. Moreover, water supply and disposal and their effects on water resources are very much site related; therefore, they do not lend themselves readily to modeling or generic solutions to issues. The planner generally is faced with balancing site-specific water issues against a broad array of other issues; the alternatives usually available to planners therefore are discussed briefly.

In conclusion, it should be noted that in water-deficient areas any additional consumptive use of water may place a significant stress on the hydrologic regime, but from the world viewpoint, water consumption is now, and will continue to be, a minor factor, even allowing for a large increase of consumption by 2000. Of more serious concern among the effects of energy development on water resources is the problem of removal of residuals generated in energy development. Chief among these problems affecting the hydrological environment are disposal of mine waters, disposal of wastes from uranium milling, discharge of degraded cooling system waters, and the special problem of disposal of radioactive materials from nuclear power plants. For a given project, these residuals and their disposal are very much site related and cannot be readily extrapolated on the basis of energy growth projections to meaningful national or world totals. The methodology for calculating effects of energy developments on water quality is still in a primitive stage, and much research will be needed before such effects can be calculated on a broad scale.

9 References

APPALACHIAN REGIONAL COMMISSION. 1969. Acid mine drainage in Appalachia. Washington, D.C., 126 p.

AXMANN, R. C. 1974. An environmental study of the Wairakei Power Plant. New Zealand Department of Scientific and Industial Research, Physics and Engineering Laboratory Report No 445, 38 p.

BUTTERMORE, P. M. 1966. Water use in the petroleum and natural gas industries. U.S. Bureau of Mines Information Circular 8284, 36 p.

BROWN, A., SCHAUER, M. I., ROWE, J.W., and HENLEY, W. 1977. Water management in oil shale mining. National Technical Information Service PB-276085.

CAMP, F. W. 1976. Processing Athabaska tar sands-tailings disposal. Canadian Journal of Chemical Engineering, vol. 55, no. 5, pp. 581-591.

CROUSE, P. C. 1981. Enhanced oil recovery: Opportunities approach reality. World Oil, vol. 193, no. 6.

DAVIS, G. H., and KILPATRICK, F. A. 1981. Water supply as a limiting factor in western energy development. Water Resources Bulletin, vol. 17, no. 1, pp. 29-35.

DAVIS, G. H., and VELIKANOV, A. L. 1979. Hydrological problems arising from the development of energy. Unesco Technical Papers in Hydrology No. 17, 32 p.

DiPIPPO, R. 1980. Geothermal energy as a source of electricity. U.S. Department of Energy Report DOE/RA/28320-1, 370 p.

EVERS, R. H. 1975. Water quality requirements for the petroleum industry. American Water Works Association Journal, February, pp. 60-67.

GIBSON, A. W. 1979. Energy development and water resources. New Zealand National Water and Soil Conservation Organization Report for National Committee for Unesco.

GULF OIL CORP.-STANDARD OIL CO.(INDIANA). 1977. Rio Blanco Oil Shale Project, revised detailed development plan Tract C-a.

HOLLYDAY, E. F., and McKENZIE, S. W. 1973. Hydrology of the formation and neutralization of acid waters draining from underground coal mines of western Maryland. Maryland Geological Survey Report of Investigations No. 20, 50 p.

HU, M., PAVLENCO, G., and ENGLESSON, R. 1978. Water consumption and costs for various steam electric power plant cooling systems. Washington, D.C., U.S. Environmental Protection Agency Report EPA 600/7-78-157.

JOHNSON, W. and MILLER, G. C. 1079. Abandoned coal-mined lands: Nature, extent, and cost of reclamation. Washington, D.C., U.S. Bureau of Mines, 29 p.

KERR, R. A. 1980. Geopressured energy fighting uphill battle. Science, vol. 207, pp. 1455-1456.

KESTIN, J. (Ed.). 1980. Sourcebook on production of electricity from geothermal energy. Washington, D.C., U.S. Department of Energy Report DOE/RA/4051-1, 997 p.

LANE, H. U. (Ed.). 1981. The world almanac and book of facts - 1982. New York, Newspaper Enterprise Association, Inc., 976 p.

MARKOS, G., and BUSH, K. J. 1980. Relationships of geochemistry of uranium mill tailings and control technology for containment of contaminants. In Proceedings 2nd U.S. DOE Environmental Control Symposium, Reston, Va., March 17-19, pp. 190-204.

MORTON, P. K. 1981. 1980 mining review. California Geology, vol. 34, no. 10.

MURRAY, C. R. and REEVES, E. B. 1977. Estimated use of water in the United States, 1977. Washington, D.C., U.S. Geological Survey Circular 765, 39 p.

OECD. 1981. Energy statistics 1975-1979. Paris, Organization for Economic Co-operation and Development Report.

OECD. 1982a. Energy balances for OECD countries 1976/1980. Paris, Organization for Economic Co-operation and Development Report, 162 p.

OECD. 1982b. World Energy Outlook. Paris, Organization for Economic Co-operation and Development Report, 473 p.

OTTS, L. E., Jr. 1963. Water requirements of the petroleum industry. Washington, D.C., U.S. Geological Survey Water Supply Paper 1330-G, 53 p.

PROBSTEIN, R. F., and GOLD, H. 1978. Water in synthetic fuel production. Cambridge, Mass., Massachusetts Institute of Technology Press, 296 p.

SCHNEIDERMAN, S. J. 1981. There's more to B C Coal's Sparwood operation than just hydromining. World Coal, vol. 7, no. 5, pp. 48-51.

SUBITZKY, S. 1976. Hydrologic studies, Allegheny County, Pennsylvania. Washington, D.C., U.S. Geological Survey Miscellaneous Field Studies Maps MF-641A-E.

UNITED NATIONS. 1982. 1980 yearbook of world energy statistics. New York, United Nations Report.

U.S. DEPARTMENT OF ENERGY. 1980. Technology characterizations--environmental information handbook. Washington, D.C., U.S. Department of Energy Report DOE/EV-0072, 219 p.

U.S. DEPARTMENT OF ENERGY. 1981a. Energy technology and the environment. Washington, D.C., U.S. Department of Energy Report DOE/EP-0026, 512 p.

U.S. DEPARTMENT OF ENERGY. 1981b. 1980 international energy annual. Washington, D.C., U.S. Department of Energy Report DOE/EIA-0219(80), 110 p.

U.S. DEPARTMENT OF ENERGY, OFFICE OF ENVIRONMENTAL ASSESSMENTS. 1980. Comparing energy technology alternatives from an environmental perspective. Washington, D.C., U.S. Department of Energy Report, 22 p.

U.S. DEPARTMENT OF THE INTERIOR. 1973. Final environmental statement for the prototype oil shale leasing program. Washington, D.C., U.S. Department of the Interior, 6 vols.

U.S. OFFICE OF TECHNOLOGY ASSESSMENT. 1980. An assessment of oil shale technologies. Washington, D.C., U.S. Office of Technology Assessment Report, 516 p.

U.S. WATER RESOURCES COUNCIL. 1968. The nation's water resources, the first national assessment. Washington, D.C., U.S. Water Resources Council, 419 p.

WORLD BANK. 1980. Energy in the developing countries. Washington, D.C., The World Bank, 92 p.

YOUNG, H. G., and THOMPSON, R. G. 1973. Forecasting water use for electric power generation. Water Resources Research, vol. 9, no. 4, pp. 800-807.

Appendix A. Glossary

ACID DRAINAGE
Waters of low pH that issue from sulfurous coal deposits; they generally have a high content of sulfuric acid, iron, and dissolved solids.

ACID RAIN
Precipitation of low pH resulting from solution of sulfur or nitrogen oxides from natural or man-made surface sources.

AEROBIC DECOMPOSITION
Decomposition of organic matter by bacteria that function in the presence of oxygen; produces water, carbon dioxide, and a small volume of residue.

AIR BLOWN
As used in coal gasification this means air is injected into the gasifier during the combustion process; if oxygen is used instead of air, a higher heat content gas is obtained.

AMINE
A derivative of ammonia in which hydrogen atoms have been replaced by radicals containing hydrogen and carbon atoms.

ANAEROBIC DECOMPOSITION
Decomposition of organic waste by bacteria that function in the absence of oxygen; properly controlled, anaerobic decomposition will produce carbon dioxide and methane and will substantially reduce the volume of waste.

BIOCHEMICAL OXYGEN DEMAND (BOD)
A measure of the amount of oxygen consumed in the biological processes that break down organic matter in water; the greater the degree of pollution, the higher the BOD.

BLOWDOWN
The continuous discharge from an evaporative cooling process required to prevent excessive buildup of dissolved-minerals content and chemical additives.

BITUMEN
Natural petroleum residues occupying spaces between individual grains of tar sands.

BREEDER REACTOR
A nuclear reactor that, in addition to generating atomic energy, creates additional fuel by producing more fissionable material that it uses.

CANDU REACTOR
Canadian deuteriated water reactor; a design employing unenriched uranium as fuel and deuteriated water as a moderator.

CATALYTIC REFORMING
A petroleum-refining process employing a catalyst in which the natural hydrocarbon molecules are restructured to increase the lower specific-gravity fractions.

CHAR
A solid organic byproduct in certain petroleum processes.

CHEMICAL OXYGEN DEMAND (COD)
A measure of the amount of an oxidant required to oxidize organic matter in water.

CLOSED-CYCLE COOLING
A cooling system in which heat is dissipated to the atmosphere on-site by cooling towers, ponds, etc.

CO_2
Carbon dioxide, a gaseous product of carbon combustion, injected into oil reservoirs in tertiary recovery.

COAL WASHING
A water bath process for reducing the non-carbonaceous content of coal.

CONSUMPTIVE USE
A use of water that decreases the quantity available because it has been evaporated, transpired, incorporated into products and crops, consumed by man or livestock, or otherwise removed from the water environment.

COOLING TOWER
Structure arranged so that water cascades down over coarse material as air is drawn through it to increase evaporative cooling effect.

CURIE
The amount of a particular radioactive material that equates to 3.7×10^{10} disintegrations per second.

DESULFURIZATION SCRUBBER
A chemical system for removing sulphur compounds from furnace exhause gases.

DEUTERIUM
H^2, the heavy stable isotope of hydrogen, atomic weight 2; used as a moderator in heavy-water reactors.

DRY-BULB TEMPERATURE
Temperature as measured by a standard thermometer in the air.

FISSION PRODUCTS
Radioactive materials formed in nuclear-fission reactions.

FLUE GAS
Exhaust gas from power-plant furnace.

FRACTIONATING TOWER
Petroleum refining equipment in which vaporized hydrocarbons are distilled into various specific-gravity fractions.

GAS COMPRESSOR
Equipment used to raise the pressure of natural gas to a level suitable for pipeline transport.

GAS SCRUBBING
A process in which water vapor and other undesirable gases are removed from natural gas.

GEOPRESSURE
A natural condition in which formation fluids are under pressures approaching the weight of the overlying materials.

GROSS WATER USE
The total water use including (1) the portions that may be returned to a water source for additional use and (2) the consumptive use.

HEAT RATE
The energy input required in a thermal process to produce one unit of energy output; expressed as kcal/kWh.

HYDROCARBON
Chemical compound containing only hydrogen and carbon.

HYDROGENATION
Any of several chemical processes in which the hydrogen content of fuels is increased.

KEROGEN
Organic material contained in oil shale; when heated to 482°C, kerogen decomposes into hydrocarbons and a carbonaceous residue.

KR^{85}
Radioactive isotope of krypton; gaseous fission product released from nuclear-power reactors.

LEACHING
Percolation of water through porous materials; in the process the water dissolves and removes soluble materials.

LIGHT-WATER REACTOR
Reactor design that employs common water as moderator and primary coolant.

LOAD OR CAPACITY FACTOR
The ratio of average load to maximum load, commonly stated as a percentage.

MAKEUP WATER
The water supplied to a water-consuming process required to maintain the needed operating volume.

METHANATION
A chemical process for adding hydrogen to a low heat-value gas so as to produce a higher heat-value product.

MISCIBLE-FLUID FLOODING
Tertiary oil recovery method in which a fluid miscible in oil is injected into the producing zone to lower the oil viscosity and displace the mixed oil toward production wells.

MUSKEG
Arctic-climate bog containing thick layers of decaying vegetation.

NUCLEAR REACTORS
The equipment systems employed for converting the thermal energy of nuclear fission reactions into other forms of energy.

OFFGAS
Gas produced through a process step, usually as a byproduct or waste.

OIL SHALE
Natural rock materials, commonly fresh-water marlstones, that contain sufficient organic matter for production of petroleum by high-temperature roasting.

ONCE-THROUGH COOLING
A heat-exchange process in which water from a river or other large body of water is allowed to flow through the system and return to the water body with the added heat load.

OVERBURDEN
Soil and rock between the land surface and a coal seam.

PHENOL
Any of a group of aromatic hydroxyl derivatives of benzene, similar in structure and composition to phenol, C_6H_5OH.

POUR POINT
The temperature at which petroleum flows freely.

PYRITE
Iron sulfide, FeS_2; principal form of inorganic sulfur in coal.

PYROLYSIS
The process of chemically decomposing a hydrocarbon by heating it in an oxygen-deficient atmosphere.

REACTOR-FUEL FABRICATION
In light-water reactor systems, the final stage of fuel preparation, in which enriched uranium oxide is formed into pellets for insertion into fuel elements.

RECYCLE WATER
Water that is collected after process use, treated, and reused at the facility.

REINJECTION
A method of disposing of waste geothermal fluids after they have given up their thermal energy.

RESIDUALS
A general term for the undesirable side effects of energy production.

RETORT
The furnace and container used to roast oil shale to drive off valuable hydrocarbons.

SHALE OIL
A multicomponent hydrocarbon mixture that results from the retorting of oil shale.

SECONDARY RECOVERY
An oil-industry practice for stimulating further production after the economic limit of production by natural flow and pumping has been reached; usually involves displacing oil toward production wells with water.

SLIMES
Suspension of fine mineral matter in water; a troublesome residual of tar-sands extraction.

SLUDGE
A general term for wet semi-solid material.

SLURRY PIPELINE
A pipeline in which coal or other solid material is transported in a water suspension.

SOLVENT REFINING
A process in which coal is dissolved and reconstituted in solid form so as to eliminate sulfur and other undesirable consitutents.

SPRAY POND
Reservoir and catch basin for water cooling system in which water is sprayed into the air to increase evaporative cooling effect.

STEAM FLOODING
A tertiary-oil-recovery method in which steam is injected into the producing zone to reduce the viscosity of the oil and displace it toward production wells.

SYNFUEL
A manufactured fuel substance produced from an energy raw material by thermochemical processes; generally cleaner and of higher heat value than raw material.

TAILINGS
Left-over material from extraction of ore; mostly rock fragments, but commonly containing some unextracted ore.

TAR SAND
Sand deposits containing bitumen in the interstitial spaces; bitumen can be recovered from tar sands by a hot washing process.

TERTIARY RECOVERY
A practice in the oil industry of stimulating additional production by employing methods of reducing the viscosity of oil, such as steam injection, or by using chemical injections for same objective.

TRACE ELEMENTS
Chemical elements at low concentrations but significant because of either toxicity or uncertainty as to effects of continued exposure.

TRANSURANIC ELEMENTS
Radioactive elements formed in nuclear-fission reactions having higher atomic numbers than uranium.

TRITIUM
H^3, the radioactive isotope of hydrogen, a gaseous fission product released from nuclear-power reactors, generally in the form of tritiated water.

TURBINE
An engine driven by the pressure of steam, water, or other fluid against the curved vanes of a wheel or wheels attached to a drive shaft.

UNIT-CONSUMPTIVE USE
Water consumption per unit of energy output of a process.

WATER FLOODING
A secondary-oil-recovery method in which water is injected into the producing zone to displace oil toward production wells.

WET-BULB TEMPERATURE
Temperature as measured by a thermometer covered with a wet cloth; temperature reading approximates that of the air at 100 per cent humidity.

ZEOLITE SOFTENERS
Natural hydrous aluminum silicates of sodium, calcium, potassium, or barium used for softening water.

Appendix B. Conversion factors

Length
1 inch (in) = 25.4 millimeters (mm)
1 foot (ft) = 0.3048 meters (m)
1 mile (mi) = 1.609 kilometers (km)

Area
1 acre (A) = 0.4047 hectares (ha)
1 ha = 10,000 m^2
100 ha = 1 km^2

Volume
1 U.S. gallon (g) = 0.0037845 cubic meters (m^3) = 3.7845 liters (l)
1 U.S. barrel (bbl, 42g)= 0.15899 m^3
1 acre-foot (AF) = 1,233 m^3
1 cubic foot (cf) = 0.028 m^3
1 cubic kilometer (km^3) = 1 x 10^9 m^3

Mass
1 short ton (2,000 lbs) = 0.9072 metric tons (t)
1 pound (lb) = 0.4535 kilograms (kg)

Flow
1 gallon per minute (gpm) = 0.003785 m^3/min
1 million gallons per day (mgd) = 1,382,289 m^3/yr
1 liter per second (l/s) = 86.4 m^3/day

Energy
1 British thermal unit (Btu) = 0.252 kilogram calories (kcal)
1 Btu = 0.000293 kilowatt-hours (kWh)
1 Btu per cubic foot (gas) = 8.89815 kcal/m^3
1 Gigawatt (GW) = 1 x 10^6 kW
1 kcal = 1.163 kWh

Water-Energy
1 gallon per million Btu = 0.01292 l/kWh
1 l/kWh = 0.000086 liters per ton oil equivalent (l/toe)

Crude oil
1 metric ton (t) = 1.16 m^3 = 7.33 bbl
1 bbl = 0.136 t
1 barrel per day (bpd) = 49.8 t/yr
1 t = 1.5 t coal as oil equivalent
1 bbl = 5.8 x 10^6 Btu
1 kWh = 0.000086 tons oil equivalent (toe)
1 ton oil equivalent = 43 x 10^6 Btu
1 bpd = 49.8 toe
1 GWh = 86 toe/yr

Temperature
$^oC = 5/9(^oF - 32)$
$1^oC = 1.8^oF$

Pressure
1 pound per square inch (psi) = 0.0703 kilograms per square centimeter (kg/cm^2)

Appendix C. Unweighted unit-consumptive-use values

Process	Unit-consumptive-use value l/toe	Unit-consumptive-use value l/kWh
Extraction		
Coal mining, surface	81	0.007
Coal mining, underground	100	0.009
Coal beneficiation	170	0.015
Oil production, secondary recovery	1,100	0.09
Oil production, tertiary recovery	4,900	0.42
Gas processing and transportation	240	0.02
Oil shale, mining and surface processing	4,700	0.40
Oil shale, modified in-situ recovery	1,300	0.11
Tar sands, mining and surface treatment	3,000	0.26
Refining		
Oil refining	1,200	0.10
Nuclear fuel cycle		
Uranium mining	9	0.001
Uranium milling	410	0.035
Uranium hexafluoride conversion	20	0.002
Enrichment, gas centrifuge	73	0.006
Enrichment, gaseous diffusion	490	0.042
Fuel fabrication	33	0.003
Fuel reprocessing	20	0.002
Steam-electric power generation[1]		
Fossil-fueled, evaporative cooling demand only	27,000	2.3
Coal-fired, total includes ash disposal and miscellaneous demands	31,000	2.7
Nuclear, light-water reactor	35,000	3.0
Geothermal, vapor-dominated systems	80,000	6.9
Geothermal, water-dominated systems	174,000	15.0
Synfuels production[2]		
Coal gasification	3,000	0.026
Coal liquefaction	1,600	0.14
Solid-fuel production	920	0.08

1 Assumes closed cooling system with cooling towers, thermal efficiency as follows: fossil fueled, 36%, nuclear, 31%, geothermal, vapor-dominated systems, 15%, water-dominated, 7-9%.

2 Assumes intermediate wet-cooling demand.

Index

Titles in this series

1. The use of analog and digital computers in hydrology. Proceedings of the Tucson Symposium, June 1966 /L'utilisation des calculatrices analogiques et des ordinateurs en hydrologie : Actes du colloque de Tucson, juin 1966. Vol. 1 et 2. *Co-edition IASH-Unesco /Coédition AIHS-Unesco.*
2. Water in the unsaturated zone. Proceedings of the Wageningen Symposium, August 1967 / L'eau dans la zone non saturée: Actes dy symposium de Wageningen, août 1967. Edited by / Edité par P. E. Rijtema & H. Wassink. Vol. 1 et 2. *Co-edition IASH-Unesco / Coédition AIHS-Unesco.*
3. Floods and their computation. Proceedings of the Leningrad Symposium, August 1967 / Les crues et leur évaluation : Actes du colloque de Leningrad, août 1967. Vol. 1 et 2. *Co-edition IASH-Unesco-WMO / Coédition AIHS-Unesco-OMM.*
4. Representative and experimental basins. An international guide for research and practice. Edited by C. Toebes and V. Ouryvaev. *Published by Unesco.* (Will also appear in Russian and Spanish) / Les bassins représentatifs et expérimentaux: Guide international des pratiques en matière de recherche. Publié sous la direction de C. Toebes et V. Ouryvaev. *Publié par l'Unesco.* (A paraître également en espagnol et en russe).
5. Discharge of selected rivers of the world / Débit de certains cours d'eau du monde / Caudal de algunos ríos del mundo / Расходы воды избранных рек мира. *Published by Unesco / Publié par l'Unesco.*
   - Vol. I : General and regime characteristics of stations selected / Vol. I : Caractéristiques générales et caractéristiques du régime des stations choisies / Vol. I : Características generales y características del régimen de las estaciones seleccionadas / Том I: Общие и режимные характеристики избранных станций.
   - Vol. II : Monthly and annual discharges recorded at various selected stations (from start of observations up to 1964) / Vol. II : Débits mensuels et annuels enregistrés en diverses stations sélectionnées (de l'origine des observations à l'année 1964) / Vol. II : Caudales mensuales y anuales registrados en diversas estaciones seleccionadas (desde el comienzo de las observaciones hasta el año 1964) / Том II: Месячные и годовые расходы воды, зарегистрированные различными избранными станциями (с начала наблюдений до 1964 года).
   - Vol. III : Mean monthly and extreme discharges (1965-1969) / Vol. III : Débits mensuels moyens et débits extrêmes (1965-1969) / Vol. III : Caudales mensuales medianos y caudales extremos (1965-1969) / Том III: Средне-месячные и экстремальные расходы (1965-1969 гг.).
   - Vol. III (part II) : Mean monthly and extreme discharges (1969-1972) / Vol. III (partie II) : Débits mensuels moyens et débits extrêmes (1969-1972) / Vol. III (parte II) : Caudales mensuales medianos y caudales extremos (1969-1972) / Том III (часть II): Средне-месячные и экстремальные расходы (1969-1972 гг.).
   - Vol.III (part III) : Mean monthly and extreme discharges (1972-1975) (English, French, Spanish, Russian).
6. List of International Hydrological Decade Stations of the world / Liste des stations de la Décennie hydrologique internationale existant dans le monde / Lista de las estaciones del Decenio Hidrológico Internacional del mundo / Список станций международного гидрологического десятилетия земного шара. *Published by Unesco / Publié par l'Unesco.*
7. Ground-water studies. An international guide for practice. Edited by R. Brown, J. Ineson, V. Konoplyantzev and V. Kovalevski. (4 supplements published).
8. Land subsidence. Proceedings of the Tokyo Symposium, September 1969 / Affaissement du sol : Actes du colloque de Tokyo, septembre 1969. Vol. 1 et 2. *Co-edition IASH-Unesco / Coédition AIHS-Unesco.*
9. Hydrology of deltas. Proceedings of the Bucharest Symposium, May 1969 / Hydrologie des deltas : Actes du colloque de Bucarest, mai 1969. Vol. 1 et 2. *Co-edition IASH-Unesco / Coédition AIHS-Unesco.*
10. Status and trends of research in hydrology / Bilan et tendances de la recherche en hydrologie. *Published by Unesco / Publié par l'Unesco.*
11. World water balance. Proceedings of the Reading Symposium, July 1970 / Bilan hydrique mondial : Actes du colloque de Reading, juillet 1970. Vol. 1-3. *Co-edition IAHS-Unesco-WMO / Coédition AIHS-Unesco-OMM.*
12. Research on representative and experimental basins. Proceedings of the Wellington (New Zealand) Symposium, December 1970 / Recherches sur les bassins représentatifs et expérimentaux : Actes du colloque de Wellington (N.Z.), décembre 1970. *Co-edition IASH-Unesco / Coédition AIHS-Unesco.*
13. Hydrometry : Proceedings of the Koblenz Symposium, September 1970 / Hydrométrie : Actes du colloque de Coblence, septembre 1970. *Co-edition IAHS-Unesco-WMO / Coédition AIHS-Unesco-OMM.*
14. Hydrologic information systems. *Co-edition Unesco-WMO.*
15. Mathematical models in hydrology : Proceedings of the Warsaw Symposium, July 1971 / Les modèles mathématiques en hydrologie : Actes du colloque de Varsovie, juillet 1971. Vol. 1-3. *Co-edition IAHS-Unesco-WMO.*
16. Design of water resources projects with inadequate data : Proceedings of the Madrid Symposium, June 1973 / Elaboration des projets d'utilisation des ressources en eau sans données suffisantes : Actes du colloque de Madrid, juin 1973. Vol. 1-3. *Co-edition IAHS-Unesco-WMO / Coédition AIHS-Unesco-OMM.*
17. Methods for water balance computations. An international guide for research and practice.
18. Hydrological effects of urbanization. Report of the Sub-group on the Effects of Urbanization on the Hydrological Environment.
19. Hydrology of marsh-ridden areas. Proceedings of the Minsk Symposium, June 1972.
20. Hydrological maps. *Co-edition Unesco-WMO.*
21. World catalogue of very large floods / Répertoire mondial des très fortes crues / Catálogo mundial de grandes crecidas / Всемирный каталог больших наводков.
22. Floodflow computation. Methods compiled from world experience.
23. Guidebook on water quality surveys.

24. Effects of urbanization and industrialization on the hydrological regime and on water quality. Proceedings of the Amsterdam Symposium, October 1977, convened by Unesco and organized by Unesco and the Netherlands National Committee for the IHP in co-operation with IAHS / Effets de l'urbanisation et de l'industrialisation sur le régime hydrologique et sur la qualité de l'eau. Actes du colloque d'Amsterdam, octobre 1977, convoqué par l'Unesco et organisé par l'Unesco et le Comité national des Pays-Bas pour le PHI en coopération avec l'AISH.
25. World water balance and water resources of the earth.
26. Impact of urbanization and industrialization on water resources planning and management.
27. Socio-economic aspects of urban hydrology.
28. Casebook of methods of computation of quantitative changes in the hydrological regime of river basins due to human activities.
29. Surface water and groundwater interaction.
30. Aquifer contamination and protection.
31. Methods of computation of the water balance of large lakes and reservoirs. Vol. I : Methodology. Vol. II : Case studies.
32. Application of results from representative and experimental basins.
33. Groundwater in hard rocks.
34. Groundwater Models. Vol. I : Concepts, problems and methods of analysis with examples of their application.
35. Sedimentation Problems in River Basins.
36. Methods of computation of low stream flow.
37. Proceedings of the Leningrad Symposium on specific aspects of hydrological computations for water projects (Russian).
38. Methods of hydrological computations for water projects.
39. Hydrological aspects of drought. (In preparation).
40. Guidebook to studies of land subsidence due to groundwater withdrawal.
41. Guide to the hydrology of carbonate rocks.
42. Water and energy: demand and effects.